Università degli Studi di Napoli "Federico II"

Scuola Politecnica e delle Scienze di Base
Dipartimento di Ingegneria Industriale

XXXI Ciclo

Tesi di Dottorato

Thermo-Fluid-Dynamics of Impinging Swirling Jets

Mattia Contino

Tutori
Prof. Ing. Gennaro Cardone
Dott. Ing. Carlo Salvatore Greco

Coordinatore
Prof. Ing. Michele Grassi

Dicembre 2018

Title | Thermo-fluid-dynamics of impinging swirling jets
Author | Mattia Contino
ISBN | 978-88-31607-38-4

Youcanprint
Via Marco Biagi 6 - 73100 Lecce
www.youcanprint.it
info@youcanprint.it

To my parents,
to my grandparents.

I do not know what I may appear to the world,
but to myself I seem to have been only like a boy
playing on the seashore,
and diverting myself in now and then finding a smoother pebble
or a prettier shell than ordinary,
whilst the great ocean of truth lay all undiscovered before me.

Isaac Newton

Contents

Contents

List of symbols

Acronyms

2D	Two-dimensional
3D	Three-dimensional
4D	Time-resolved three-dimensional
ART	Algebraic reconstruction technique
C-SMART	Camera simultaneous multiplicative algebraic reconstruction technique
CCD	Charge-coupled devices
CMOS	Complementary metal-oxide-semiconductor
DNS	Direct Numerical Simulation
EPOD	Extended Proper Orthogonal Decomposition
FPA	Focal plane array
IPR	Iterative particle reconstruction
IR	Infrared
ISV	Inner secondary vortex
IT	Integration time
K-L	Karhunen–Loève
LDV	Laser Doppler Velocimetry
LIF	Laser-Induced Fluorescence
LOS	Line of sight
LWIR	Long-Wave Infrared
MART	Multiplicative algebraic reconstruction technique
MTE	Motion tracking-enhanced
MWIR	Mid-Wave Infrared

Nd:YAG	Laser category which employs a Neodymium-doped Yttrium Aluminium Garnet crystal as lasing medium for solid-state lasers
Nd:YLF	Laser category which employs a Neodymium-doped Yttrium Lithium Fluoride crystal as lasing medium for arc lamp-pumped and diode-pumped solid-state lasers
NETD	Noise equivalent temperature difference
NWIR/SWIR	Near-/Small-Wave Infrared
OSV	Outer secondary vortex
PCA	Principal Component Analysis
PID	Proportional–integral–derivative
PIV	Particle Image Velocimetry
PMMA	Polymethyl methacrylate
POD	Proper Orthogonal Decomposition
PTV	Particle Tracking Velocimetry
PVC	Precessing vortex core
PWM	Pulse-width modulation
RMS	Root-mean-square
RTD	Resistance temperature detector
SMART	Simultaneous multiplicative algebraic reconstruction technique
SMTE	Sequential motion tracking-enhanced
SR	Vortex filament of ω_r
STB	Shake-the-box (Schanz et al., 2016; Schröder et al., 2015)
SVD	Singular Values Decomposition
T-PIV	Tomographic Particle Image Velocimetry
TR T-PIV	Time-resolved Tomographic Particle Image Velocimetry
UV	Ultra-Violet

Roman Letters

A, B	Generic data sets
C_ρ	Correlation coefficient

C_s	Two-point spatial correlation matrix
C_t	Two-point temporal correlation matrix
D	Diameter of the nozzle
d_{diff}	Particle diameter related to diffraction effects
d_{geom}	Particle diameter related to geometric effects
D_h	Hub diameter of a swirling nozzle with helical pattern
d_p	Physical particle diameter
d_{pix}	Pixel camera size in physical units
d_τ	Diameter of particles on the camera sensor
d_τ^*	Diameter of particles on the camera sensor expressed in pixel units, $d_\tau^* = d_\tau/d_{\mathrm{pix}}$
E	Three-dimensional intensity distribution
$\mathscr{F}$	Three-dimensional vector field
F	Matrix of snapshots
f	Fluctuating part of the three-dimensional vector field denoted by $\mathscr{F}$
$f_\#$	f-number, equal to the ratio of lens focal length to aperture diameter
f_P	PIV sub-system acquisition frequency
f_T	Thermographic sub-system acquisition frequency
G_θ	Axial flux of swirl momentum
G_x	Axial flux of axial momentum
H	Distance of the nozzle from the impingement surface
I	Projection of the three-dimensional intensity distribution on the camera image plane
$K_1,\ K_2$	First and second universal constants
L_z	Length of epipolar line
M	Magnification factor, $M = d_{\mathrm{geom}}/d_p$
m	Number of voxels
M_0	Source value of the streamwise momentum
N	Number of cameras
n	Number of pixels

N_g	Number of ghost particles
n_m	Number of modes
N_{obs}	Number of observations
N_p	Number of true particles
n_p	Measurement points
N_S	Source density, $ppp \cdot \left(\frac{\pi d_\tau^*}{4} \right)$
n_t	Number of temporal realisations
O	Origin of the reference system
P	Energy flux/energy rate per unit body area; Time-averaged static pressure
p	Static pressure
ppp	Particle per pixel, average particle area in pixels
Q	Q-criterion vortex identification (Hunt et al., 1988)
Q_f	Reconstruction quality factor (Elsinga et al., 2006)
$\mathbb{R}$	Set of real numbers
r	Radial direction in the cylindrical coordinate system adopted; Rank of a matrix
$r_{1/2}$	Jet half-radius
Re	Reynolds number
$R(\underline{s})$	Cross-correlation function
S	Swirl number
$\underline{s}$	Vector of shifts
S_{cr}	Critical value of swirl number (Billant et al., 1998)
St	Strouhal number
T	Temperature
t	Time coordinate
t_1, t_2, t_3, t_4	Time delays in the experimental apparatus synchronisation settings
T_C	Low reference temperature
T_{cal}	Temperature after linear background correction procedure
T_{EndTherm}	Temperature in the test tank at the end of the thermal

	acquisition
T_H	High reference temperature
T_{Init}	Temperature in the test tank at the beginning of the experimental test
T_{InitPIV}	Temperature in the test tank at the beginning of the PIV acquisition
$T_{\text{InitTherm}}$	Temperature in the test tank at the beginning of the thermal acquisition
$T_{\text{not cal}}$	Temperature map before the application of the linear background correction
T_{obs}	Time of observation
U, V, W	Time-averaged axial, radial and azimuthal components of velocity in a cylindrical reference system
u, v, w	Axial, radial and azimuthal components of velocity in a cylindrical reference system
U_0	Bulk velocity, reference velocity
U_c	Centreline velocity of a free round jet
$U_{r_{1/2}}$	Velocity at radial distance $r = 0.5D$
$V_{\parallel}$	Norm of velocity components parallel to the selected plane
w_{vort}	Analytical distribution of azimuthal component of velocity in the Rankine vortex model
x	First direction of Cartesian and cylindrical coordinate systems adopted
$\underline{x}$	Volume index; space coordinate
X_c, Y_c	Camera reference system
x_c, y_c, z_c	World reference system
y, z	Remaining directions of the laevorotatory Cartesian coordinate systems adopted

Greek Letters

α_r	Radiation fraction absorbed by the body
β	Stefan-Boltzmann constant

χ_r	Radiation fraction reflected by the body
Δ_{field}	Extension of the in-focus region, referred to as "depth of field"
δ_{ij}	Kronecker delta symbol
ΔT	Temperature difference
Δt	Time increment
δ_t	Local time-averaged diameter of the jet
ϵ	spectral emissivity coefficient
ε_r	Search radius in the particle matching step included in the self-calibration procedure
η	Non-dimensionalised wall temperature
Γ	Circulation
Λ	Diagonal matrix of the eigenvalues
λ	Wavelength; Eigenvalue
λ^*	Wavelength at which a black body emits the maximum emissive power
μ	Relaxation parameter
Ω	Interrogation volume
ω_{ij}	Weighting function which specify the contribution of each jth voxel to the ith pixel
Ω_{obs}	Volume of observation
$\omega_x, \omega_r, \omega_\theta$	Normalised axial, radial and azimuthal vorticity component
Φ	Time-average of a generic (three-dimensional) statistical independent distribution; Matrix of spatial decomposition basis of the fluctuating field
ϕ	Generic (three-dimensional) statistical independent distribution; Spatial decomposition basis of the fluctuating field
φ	Characteristic parameter of a swirling nozzle with helical pattern
Ψ	Time-average of a generic (two-dimensional) statistical

	independent distribution; Matrix of temporal decomposition basis of the fluctuating field
ψ	Generic (two-dimensional) statistical independent distribution; Temporal decomposition basis of the fluctuating field
ρ	Density of fluid under investigation
σ	Standard deviation
σ_G	Standard deviation of the Gaussian filter
τ_r	Radiation fraction transmitted by the body
θ	Azimuthal direction in the cylindrical coordinate system adopted
ϑ	Swirl generator vane angle
ξ	Mean value of the interrogation volume

Apices

$(\cdot)'$	Fluctuating contribution of the investigated quantity
$\overline{(\cdot)}$	Time-average
$\langle\cdot\rangle$	Ensemble average
N	Investigation volume in proximity of the nozzle
W	Investigation volume in proximity of the wall

Pedices

∞	Asymptotic value
A, B	Quantity related to two generic data sets
b	Black body
e	Quantity related to Extended POD modes
HB	Second boundary of the investigation volume
LB	First boundary of the investigation volume
r	Real object; radial component
ref	Reference value

Summary

The superimposition of a tangential motion on a conventional round jet has been demonstrated to significantly affect the large-scale topology of the flow. Swirling flows are widely employed, in particular in the impinging configuration, in several industrial processes which involve both non-reacting and reacting applications. In the former case, they are typically used in drying, tempering and spray generation. In the latter, they are adopted to reduce pollutant emissions, to stabilise the flame and to finely tune its size.

In the present dissertation the simultaneously acquired thermal and three-dimensional velocity fields of an impinging hot jet emerging from a custom swirl generator in a cold ambient are presented. The velocity and temperature fields are experimentally measured using time-resolved Tomographic PIV and high-speed Infrared (IR) thermography in a combined system.

The detailed description of a custom swirl generator is given. The assembly, made of three 3D-printed objects, is able to generate swirl numbers ranging from zero to 0.74. Such a device is placed in a simultaneous measurement system equipped with temperature control systems, a flow meter and a visible-IR transparent optical access which acts also as impingement wall.

The time-averaged velocity profiles of a free swirling flow are discussed and the swirl number related to each swirl generator is evaluated. Five swirl generators are selected to discuss the effect of low to high swirl numbers on the instantaneous three-dimensional dynamics in proximity of the nozzle. The vortical structures related to a swirling jet are recognised in the presented snapshots. Then, the main features of the free flow field are extracted applying Proper Orthogonal Decomposition (POD) technique.

The impinging configuration is investigated simultaneously acquiring the three-dimensional velocity field and the temperature footprint of the hot jet exhausting in a cold ambient. Time-averaged velocity and temperature profiles are estimated in proximity of the wall. Then the time-dependent features of both quantities are discussed. The thermal fluctuations dynamics is analysed extracting the time sequences at three different distances from the impingement centre. The unsteady velocity field behaviour is discussed, comparing the instantaneous vortical structures with the temperature fluctuations field.

In order to extract the main features of an impinging swirling jet, POD analysis is applied to three-dimensional velocity fluctuations and to thermal fluctuations. The most energetic modes are discussed for both quantities.

Taking advantage of synchronisation and of knowledge of relative positioning of thermal and velocity frames, two different correlation techniques are applied and their outcomes are discussed. The Extended POD technique is applied to verify a global spatio-temporal correlation. Then, the time-averaged correlation map between the lowest slice of the velocity three-dimensional fluctuations distribution and the thermal footprint is discussed.

Keywords: impinging swirling jet, time-resolved Particle Image Velocimetry, high-speed Infrared thermography, velocity-temperature correlation, passive scalar transportation, mixing.

1 | Introduction

Impinging jets are still an attractive investigation topic because of their notorious high heat and mass transfer rate capabilities. They are typically used in a variety of industrial processes like electronic chip cooling (Pavlova and Amitay, 2006), food treatment (Moreira, 2001) and drying (Kurnia et al., 2017; Mujumdar, 2014).

One of the major drawbacks of conventional jets is the non-uniform heat transfer distribution (Carlomagno and Ianiro, 2014). The superimposition of tangential motion to a conventional jet is one of the most effective methods to fix this issue and to increase the heat transfer rate. Swirling flows are of fundamental relevance in sprays (Zeng et al., 2015) and combustion and flame stabilisation (Candel et al., 2014). The free flow field of a swirling jet is well known in literature (Alekseenko et al., 2018; Graftieaux et al., 2001) but there are only few details about the unsteady three-dimensional large scale structures behaviour involved in the impingement configuration.

Most experimental studies on impinging jets are conducted in isothermal environment. In case of temperature difference between the jet and the ambient, the entrainment phenomenon acts a crucial role in the impinging jet heat transfer performance (Baughn et al., 1991; Goldstein et al., 1990; Viskanta, 1993). As long as temperature difference is small, it does not affect the fluid motion. Hence, the transport of temperature fluctuations intended as passive scalar can be fairly used as tracer of jet impingement dynamics (Warhaft, 2000). Hence, the simultaneous measurement of both three-dimensional velocity field and temperature footprint of a swirling impinging jet would be extremely interesting.

In the present section the main literature review about the flow field of free and impinging jets will be presented. The circular jet will be taken as reference case and compared to the behaviour of a swirling jet. A brief introduction to the role of mixing and transport of passive scalars on heat transfer of an impinging jet will be presented. In the last section the motivation of the work will be defined.

1.1   Submerged round jets

A *jet* is a stream of liquid or gas produced by a pressure drop through an orifice (List, 1982). Such flows can be recognised in two main boundary conditions, depending on whether the jet exhausts in the same medium or not. A detailed compendium of such kind of flows is reported in Rajaratnam (1976) and Abramovich et al. (1984).

It is out of the scope of this work to review all cases in which submerged jets occur. In the following, the basic concepts dealing with free and impinging submerged jets will be presented. The effect of superimposed swirl motion will also be discussed.

1.1.1   Free submerged round jets

Let us consider a circular submerged jet of Newtonian fluid emerging from a nozzle of diameter D in a large stagnating chamber (figure 1.1). Even though the downstream development depends on both initial and boundary conditions (Bradshaw, 1966; Xu and Antonia, 2002), it is fair to define a polar co-ordinate system (x, r, θ) with x direction aligned with the nozzle axis. The characteristic velocity of the jet, the *bulk velocity*, is referred to as U_0. The centreline velocity is named U_c which is also the maximum velocity.

Two are the characteristic lengths involved directly or indirectly in all literature results about free jets: the jet half-radius $r_{1/2}$, which is

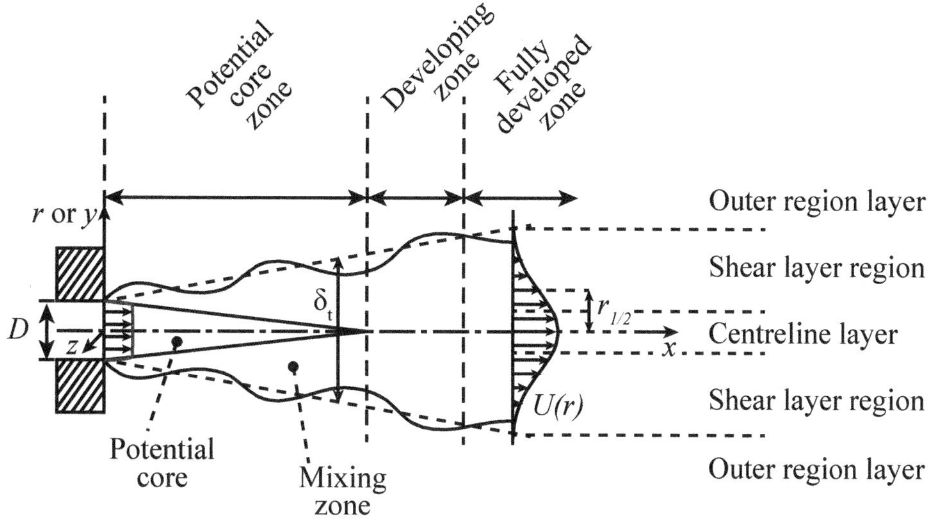

Figure 1.1: Schematic of main regions involved in a free submerged jet (Ball et al., 2012).

determined by $U_{r_{1/2}} = U_c/2$, and the local time-averaged diameter of the jet δ_t.

The flow field of a free submerged jet is characterised by three main regions: the near field, the intermediate field and the far field. In the nearby of the nozzle a conical region with undiminished velocity equal to U_0 is noticeable. This region is called *potential core*, it typically extends within the range $0 \leq x/D \leq 7$ (Ball et al., 2012). The transition or developing region is included between $x/D = 7$ and $x/D = 70$. In this region anisotropic turbulent structures form and develop. Typically this region is characterised by instabilities which are strongly dependent on the Reynolds number. The far field is located approximately at $x/D \geq 70$ (Ball et al., 2012). In this region the flow is considered fully developed or in equilibrium. Mathematically speaking, this means that all terms of governing equations are in the same relative balance independently from the x coordinate (George, 1989).

The inspection along the radial direction of the flow field of a free jet leads to consider three concentric layers: the centreline zone, the shear layer and the outer layer (Ball et al., 2012). In the centreline region, the

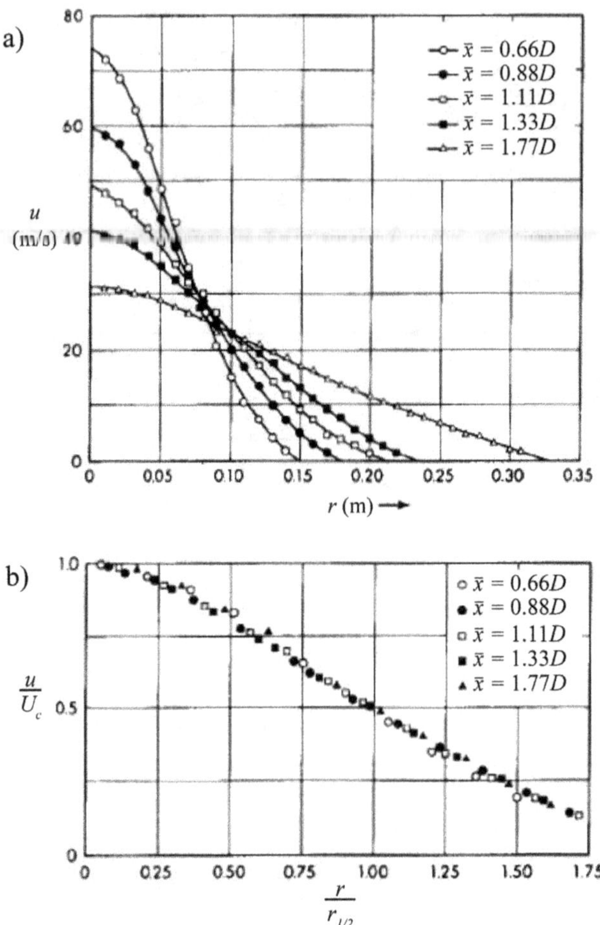

Figure 1.2: Velocity distributions of a free submerged jet at several distances from the nozzle exit (Rajaratnam, 1976). a) Distribution of the axial velocity along the radius. b) Velocity profiles in dimensionless quantities.

mean velocity of the flow is maximum. The direction of velocity vector is basically aligned to jet axis. In the shear layer, the interaction between the flow and the quiescent fluid causes the formation of large vortex cores. The outer layer refers to the undisturbed region in proximity of the jet.

The first steps of research towards characterisation of free jets flow field were driven by the need to build a reliable statistics database of the velocity fields of turbulent jets (Labus and Symons, 1972).

The simplest description of the overall behaviour is given by

Abramovich et al. (1984) who, reviewing the experimental data of Trüpel (1914), states: *The velocity profile becomes "lower" and "wider" with increasing distance from the beginning of the jet.* This conclusion can be deduced from the representation of velocity in cylindrical coordinate. As a matter, moving away from the nozzle, any section experiences a velocity decreasing and a spreading of the velocity profile (figure 1.2a). A similarity condition can be verified reporting on a single coordinate plane the ratio between u and U_c versus r and $r_{1/2}$ (figure 1.2b).

As it can be noticed, the above description is suitable to provide a full description of the flow field but it does not take into account what are the involved flow features. Fiedler (1988) underlines the role of coherent structures in turbulent flows. Three equally distributed preferred modes exist in axisymmetric flows: the ring, the single helix and the double helix (figure 1.3). The investigation of the interaction between such vortical structures is crucial to get deeper in the phenomena correlated to such flows: basic transport, mixing and noise generation phenomena (Hussain, 1986).

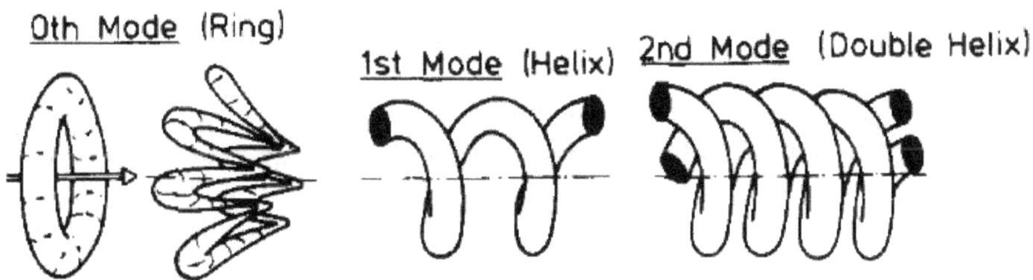

Figure 1.3: Schematic structure of a transitional free submerged jet (Fiedler, 1988).

Vortical structures are mainly located in the developing zone. As it can be noticed, the interface edges of the schematic in figure 1.1 are highly convoluted. They represent the shear layer, or mixing interface, between the jet flow and the surrounding fluid at rest (Davidson, 2015). As a matter, the flow field in the nearby of the boundary layer does not travel undisturbed along the jet extension but it is affected by instabilities. It is worth noting that, even if the momentum flux in a jet is conserved, the mass flux is not necessarily. Hence, a turbulent fluid flow drags ambient fluid toward the jet axis increasing its mass flux as much as the distance from the jet origin grows itself. This process known as *entrainment* is triggered by the convoluted outer boundary layer. In figure 1.4

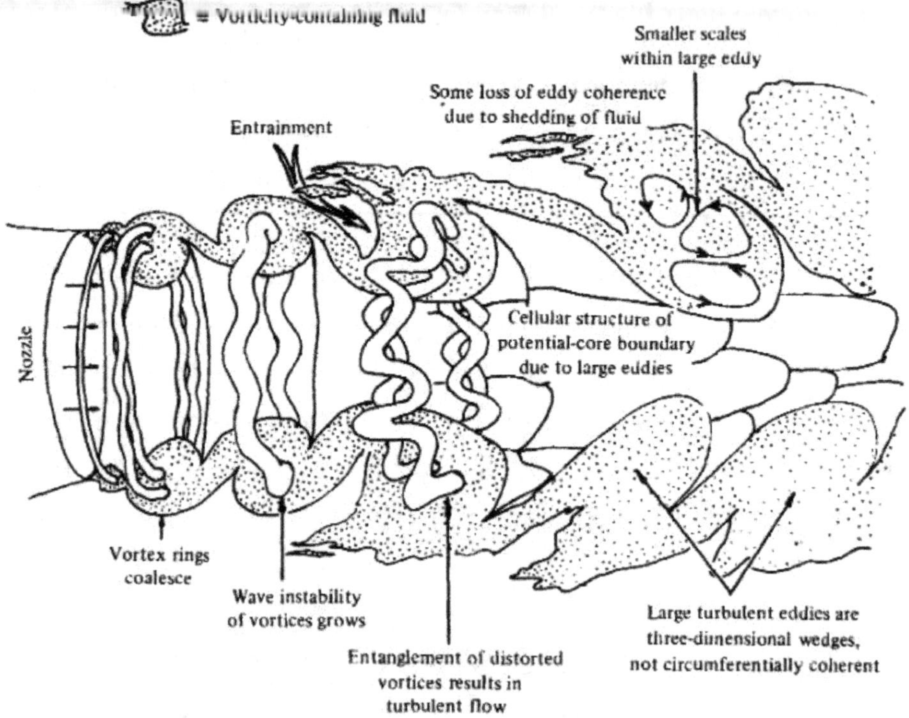

Figure 1.4: Schematic structure of a transitional free submerged jet (Yule, 1978).

the schematic of the main phenomena involved in a transitional round jet is reported. As it can be noticed, the instability of the shear layer produces ring-shaped vortices which in turn cause entrainment. As the rings move downstream they generally coalesce with the neighbouring rings thus increasing their scale. Furthermore, the gradual increasing of axial fluctuations with distance from the nozzle causes the deformation of the vortex rings.

An experimental visualisation of such phenomena is reported in figure 1.5. The filament line patterns of a turbulent jet, until the first two diameters, are clearly visible. Even in its simplicity, this smoke-wire flow technique provides significant information. The dyed flow is clearly distinguishable from the fluid at rest, so the shear layer is easily detectable. The shape of the shear layer is heavily convoluted which means it is intrinsically unstable. In the meanwhile, the vortical lump is convecting downstream, it grows in size and experiences a roll up phenomenon. Such vortices form at small distances from the nozzle and they are convected at significant distance from the nozzle.

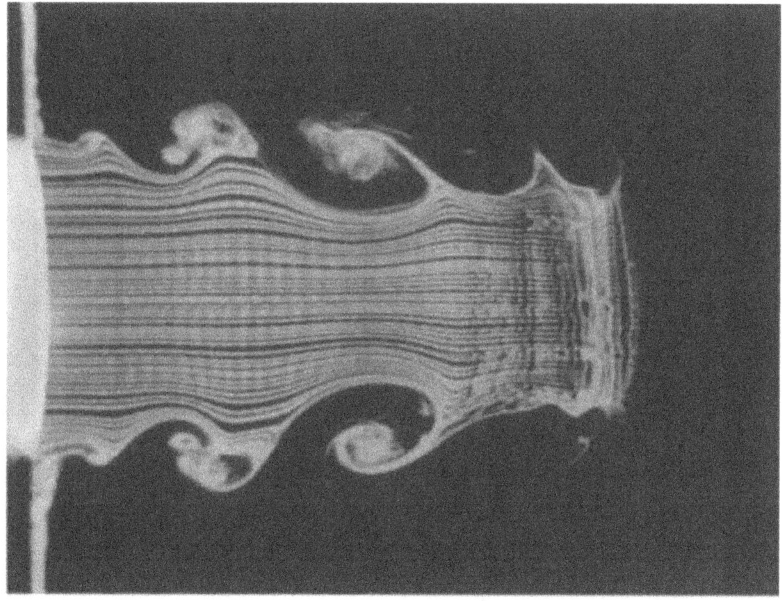

Figure 1.5: Visualisation of the instability of a free turbulent jet at $Re = 10^4$ (Popiel and Trass, 1991).

The flow field dynamics of a free jet can be described referring to two kinds of instabilities: the primary and the three-dimensional instabilities (Liepmann and Gharib, 1992). The formers are related to the primary flow. The vortices reported in figure 1.5 are associated with these primary instabilities. Once several vortices populate the flow field, if two of them get close enough, they merge. Typically, this kind of vortices is generated by the instabilities originated from the nozzle mouth. The merging phenomenon continues downstream until the vortical structures break abruptly. The breakdown typically coincides with the end of the visible potential core and the start of mixing transition (figure 1.6). As a

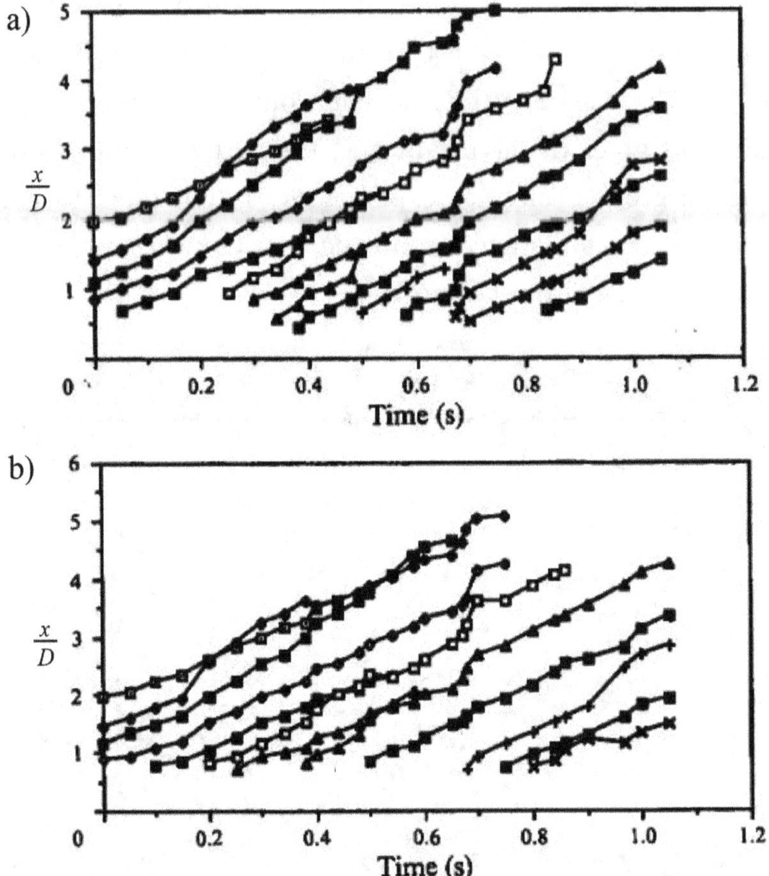

Figure 1.6: Axial position of vortex cores versus time (Liepmann and Gharib, 1992). a) Dynamics of upper vortex cores. b) Dynamics of lower vortex cores.

matter, the upper and lower vortex cores are symmetric until $x/D < 2$. Then the rings become unstable and begin tilting and warping.

The three-dimensional instabilities are related to a cross-wise unsteadiness. They appear to be similar to fingers extending out of the flow core as much as the distance from the nozzle increases. These structures move radially in and out with the passage of vortex rings. The technique applied by Liepmann and Gharib (1992) to spot and follow the vortex rings was the Laser-Induced Fluorescence (LIF). The technique was not suitable to provide three-dimensional information. Applying the

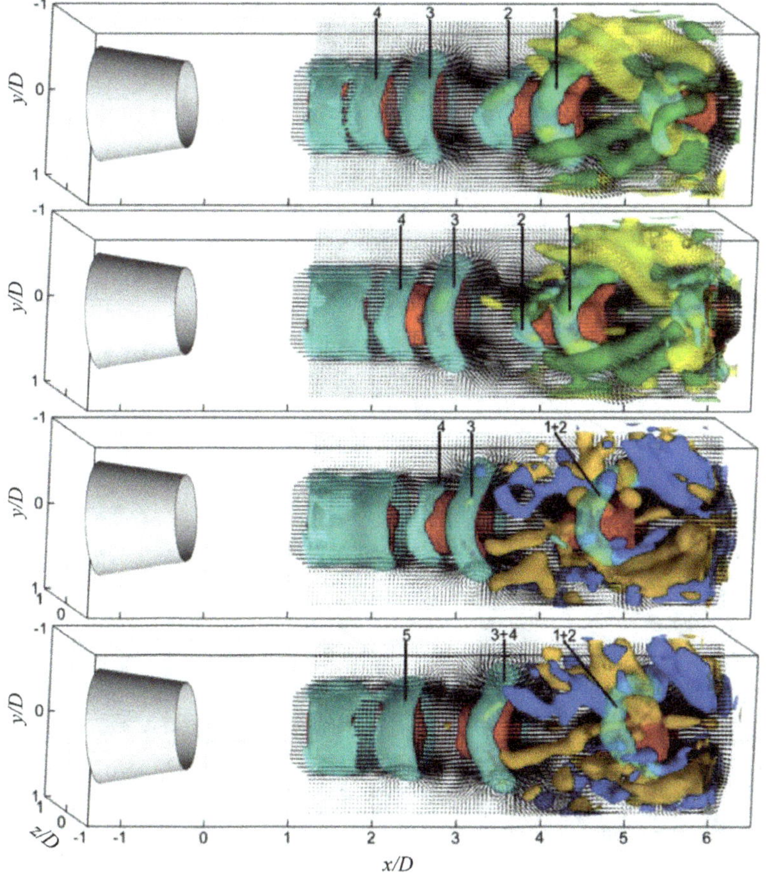

Figure 1.7: Time sequence visualisation of a circular jet, $Re = 5000$ (Violato and Scarano, 2011). Isosurfaces of axial velocity -1.05 (red), isosurface of normalised azimuthal and axial vorticity components $\omega_\theta = 4$ (cyan), $\omega_x = -1.2$ (green) and 1.2 (yellow); in the last two snapshots, isosurfaces of normalised azimuthal and radial vorticity components $\omega_\theta = 4$ (cyan), $\omega_r = -1.2$ (blue) and 1.2 (orange).

recent innovative diagnostic techniques, Violato and Scarano (2011) described the free flow of a round jet through the application of high-speed Tomographic-Particle Image Velocimetry (Elsinga et al., 2006).

The isosurface of normalised azimuthal vorticity component $\omega_\theta = 4$ identifies axisymmetric vortices labelled with progressive numbers from one to five. The snapshots report a streamwise and inward acceleration in the jet core given by the vortex 4. This velocity gradient stretches the vortex 3 and, as a consequence, the distance between the two vortices gradually decreases. The behaviour of vortex 1 and 2 is different. They are already in an advanced position along the jet development. This means that they already experienced instabilities phenomena. A crosswise sectioning of the radial vorticity further helps to spot the radial motion induced by the multiple rings. The cross-sectional plot at $x/D = 3.8$ illustrates several and not organised vortex filaments of ω_r (labelled as SR in figure 1.8b). This suggests that they are formed in the surrounding area of primary ring instabilities, the so called *braid region*.

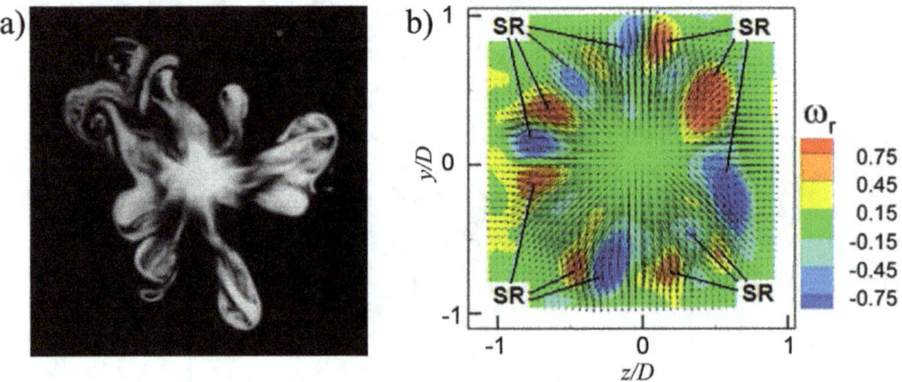

Figure 1.8: Comparison of visualisation techniques of the crosswise section in the near field of a free turbulent jet. a) Cross-sectional LIF visualisation of braids of a free turbulent jet at $x/D = 3.5$ (Liepmann and Gharib, 1992). b) Crosswise radial vorticity of a free turbulent jet at $x/D = 3.8$ (Violato and Scarano, 2011).

1.1.2   Impinging submerged round jets

The most natural extension of investigations about free round turbulent jet is the study of the interaction with obstacles or significant disturbances, like a wall. Impinging jets are well-known for their valuable heat and mass transfer rate capabilities especially at low nozzle-to-plate distances (Carlomagno and Ianiro, 2014). Together with the already cited (Abramovich et al., 1984; Jambunathan et al., 1992; Viskanta, 1993) the

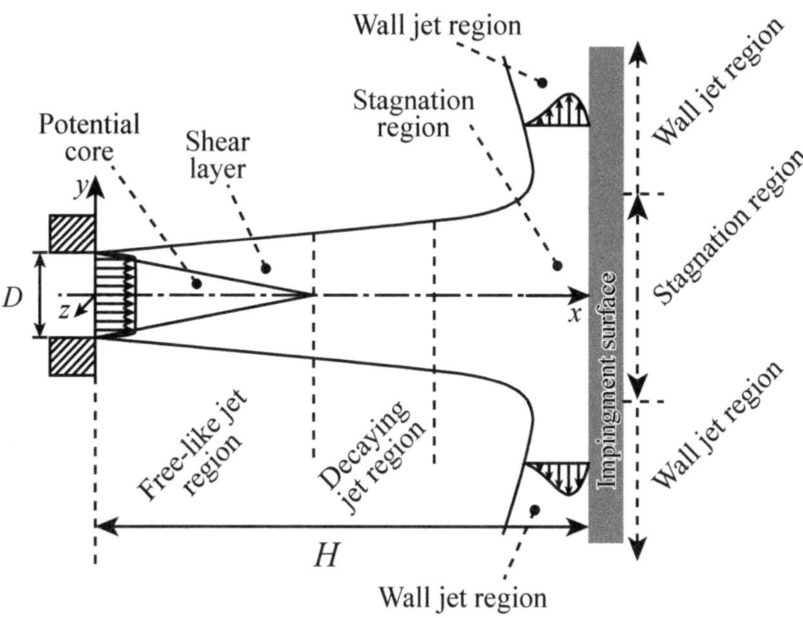

Figure 1.9: Schematic of the flow field of an impinging submerged jet.

numerical modelling done by Zuckerman and Lior (2006) is also valuable.

The flow of an impinging submerged jet experiences different steps, some of which may not exist with the decreasing of the distance between the nozzle and the impingement wall. Hence, the most general case is for large nozzle-to-plate distances.

Let us consider a turbulent jet emerging from a round nozzle and a rigid wall perpendicularly placed at a distance from the nozzle (figure 1.9). The flow field is characterised by three main regions: the potential core, the dissipation region beyond the apex of potential core, the stagnation region and the wall jet region. The first step of the jet path toward the wall is characterised by a region of peripheral velocity gradient caused by the interaction between the jet and the ambient fluid at rest through the shear layer. The jet behaves like it would be free to evolve along its axis. Proceeding downstream the flow is progressively decelerated and the velocity profile spreads in the radial direction. If the nozzle is sufficiently far from the impinging wall, a potential core still exists in the impingement configuration. Once the potential core is terminated, the developing region begins. At this stage, the flow field is substantially identical to the free configuration.

The approach to the wall provides substantial differences. The wall represents a rigid obstacle for the jet which is forced to undergo a sudden change in direction from axial to radial in a region called *quasi-stagnation region*. The actual stagnation point is located on the wall in correspondence of the jet axis. Once the velocity decreases in the nearby of the stagnation region, the flow undergoes a sudden acceleration which is favoured by a positive pressure gradient. As a matter, the stagnation region is a high static pressure region.

Once the velocity is deflected from the axial to the radial direction, the flow has space to radially spread like a semi-confined flow. The last region the flow experiences is called *wall jet region* (Gauntner et al., 1970). Hence, the horizontal component is constantly accelerated from the stagnation region to a maximum located at about one nozzle diameter from the impingement. Moving away from the nozzle along the horizontal direction, the jet grows in thickness. The shear layer is present also in this region. Such a region is influenced by both the no-slip condition at the wall, which involves a velocity gradient in proximity of the impingement plate, and the velocity difference between the wall jet and the quiescent surrounding ambient fluid. As in the free jet case, the shear layer acts as a trigger to the entrainment phenomenon. This results in the growing of wall jet thickness. In addition, the location of highest flow speed progressively shifts farther from the wall while the average peak speed decreases (figure 1.10). In the fully developed wall jet region, the maximum speed tends to zero with the increase of the distance from the impingement. An extensive survey about the flow field of an impinging round jet is reported in Gauntner et al. (1970).

It is worth noting that as the nozzle-to-plate distance is reduced, some of the reported regions no longer exist. In case the nozzle is located at distance $H \leq 2D$ from the wall, the high static pressure zone in proximity of the stagnation region acts a main role in the flow development.

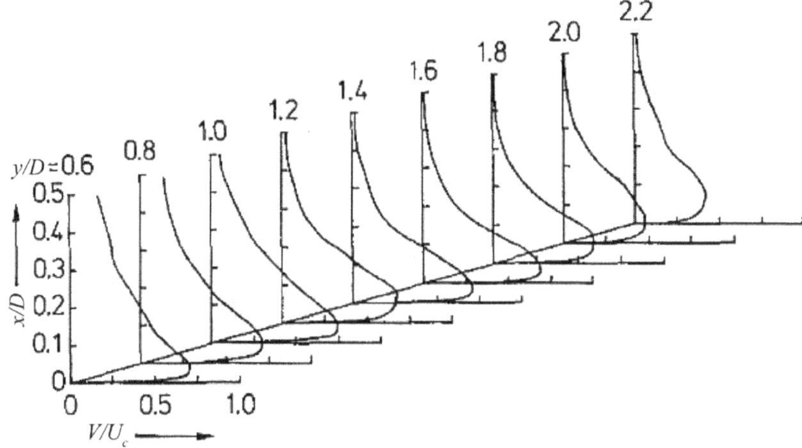

Figure 1.10: Profiles of the mean vertical velocity field at various distances y from the jet centreline (Landreth and Adrian, 1990).

Eventually, also the conical shape could vanish. As a matter, at small H/D ratios, the velocity profile of the jet does not have sufficient room to develop and the arrival velocity is essentially uniform. In that case impingement occurs within the jet potential core.

An impinging jet is considered laminar or turbulent depending on the characteristics of the flow. McNaughton and Sinclair (1966) summarised four types of jets:

- Dissipated-laminar jet, $Re < 300$;

- Fully laminar jet, $300 < Re < 1000$;

- Semi-turbulent jet, $1000 < Re < 3000$;

- Fully laminar jet, $Re > 3000$.

Figure 1.11 reports a smoke visualisation of an impinging turbulent jet. Large-scale toroidal vortices developing in the mixing region are noticeable. The approach to the wall causes vortices stretching and increasing in size. Once reached the wall, the eddies stretch and roll up moving along the wall. At radial distance of $r/D = 1$ from the impact, it is visible a transition zone where the eddies are suddenly merged and the

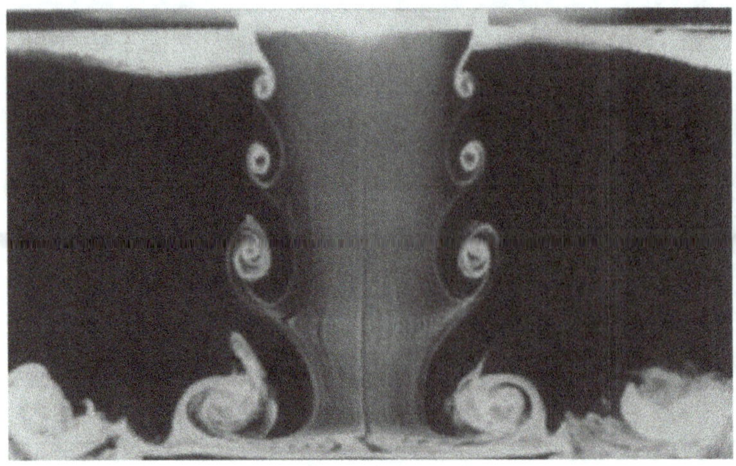

Figure 1.11: Visualisation of a round impinging turbulent jet; $Re = 3500$, $H/D = 2$ (Popiel and Trass, 1991).

flow develops in a turbulent radial wall jet. A deeper insight into the time-dependent evolution of structures of an impinging jet is provided by Violato et al. (2012). In figure 1.12 the time evolution of a round turbulent impinging jet ($Re = 5000$) is reported . The vortex rings rapidly increase in diameter along their path toward to the wall. In addition, the structures related to radial and axial vorticity are gradually tilted and stretched.

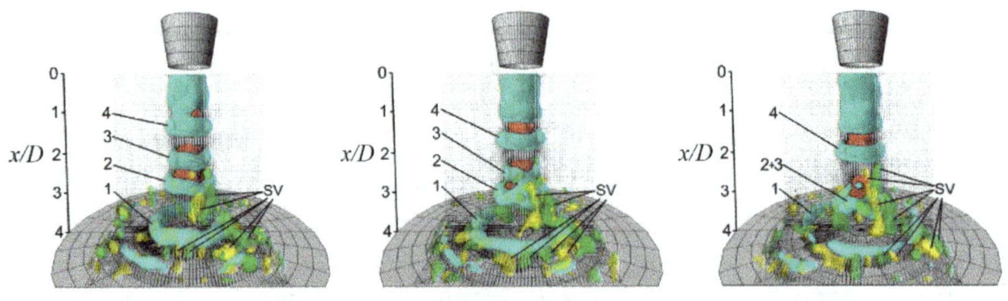

Figure 1.12: Temporal sequence of instantaneous iso-surface axial velocity 1.45 (red), azimuthal vorticity $\omega_\theta = 4$ (cyan), axial vorticity $\omega_x = -1.2$ (green) and 1.2 (yellow). Vectors on plane $y/D = 0$. $Re = 5000$ (Violato et al., 2012).

1.2 Submerged swirling jets

The addition of a tangential motion to a jet modifies its spreading rate, entrainment and mixing performances. For this reason, several industrial processes take advantage of swirling flows. They are largely applied in pumps, jet engines and vacuum cleaners. The reason for the application of rotating motions depends on the context. For inert jets, swirling jets are typically applied to take advantage of the jet growth and entrainment enhancement. As a consequence, heat and mass transfer are generally improved (Huang and El-Genk, 1998; Ianiro and Cardone, 2012). Hence, the industrial processes which could be interested in swirling jet are drying, tempering, furnace heating (Fudihara et al., 2007), food treatment (Moreira, 2001) and electronic chip cooling (Pavlova and Amitay, 2006). In case of reacting flows, swirling flows are employed to tune and control the flame size (Lilley, 1974), flame shape stability (Syred et al., 1971) and combustion intensity (Beér and Chigier, 1972; Lilley, 1977).

Another interesting branch of research on swirling flows focuses on controlling the flow field characteristics in order to match the designer needs. As a matter, good design can optimally control emissions (Micklow et al., 1993) or save energy and materials (Tsukaguchi et al., 2007). The several and specific applications offered by rotating flows suggest the critical interest about this kind of flows.

1.2.1 Generation of swirling flows

Swirling flows are typically generated by the superimposition of a rotational motion to the free flow generated by a round nozzle. All effects provided by a swirling jet strictly depend on the swirl ratio imparted to the flow. The swirling rate is measured by a dimensionless parameter, the *swirl number S*. If not differently declared, the coordinate system applied in the following will be always a cylindrical polar coordinate system. Hence, the velocity components in (x, r, θ) will be named (u, v, w) with the first direction aligned with the jet axis (figure 1.13).

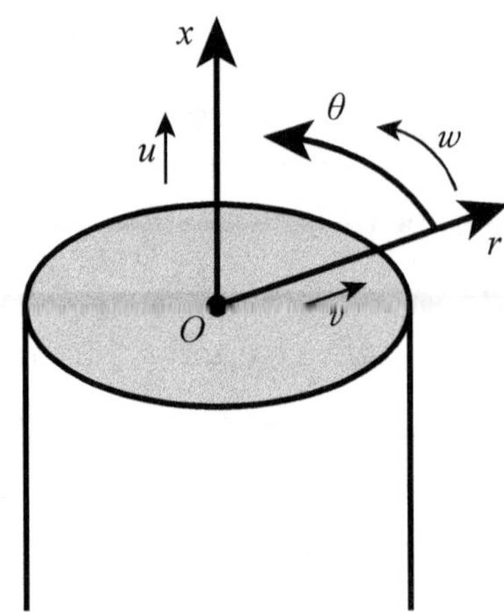

Figure 1.13: Schematic of the adopted cylindrical reference coordinate system.

The swirl number S is defined as the ratio between the axial flux of swirl momentum and the axial flux of axial momentum (Gupta et al., 1984):

$$S = \frac{G_\theta}{G_x D/2} \tag{1.1}$$

where

$$G_x = \int_0^\infty \left[\rho u^2 + \rho \overline{u'^2} + (p - p_\infty) \right] r\,dr \tag{1.2}$$

is the axial flux of axial momentum, and

$$G_\theta = \int_0^\infty (\rho uw + \rho \overline{u'w'}) r^2 dr \tag{1.3}$$

is the axial flux of the swirl momentum and $D/2$ is the nozzle radius. In equations (1.2) and (1.3) r represents the radial coordinate used as integration variable, p is the static pressure and p_∞ is the asymptotic static pressure. Four ranges of swirl number are commonly defined: non-swirling jets ($S = 0$), weakly swirling jets ($0 < S \leq 0.4$), moderate swirling jet ($0.4 \leq S \leq 0.6$) and strongly swirling jets ($S > 0.6$). The definition provided in equation (1.1) is difficult to use in practical applications. This is due to the hardness of directly measuring p in each point of the flow field.

A swirling flow can be obtained in several ways:

1. tangential entries;

2. guided vanes;

3. direct rotation.

In case of an axial-plus-tangential entry swirl generator, two flows are finely controlled in mass flow rate (figure 1.14). The tuning of the swirling rate can be obtained adjusting the ratio between the axial and the tangential flow rates. Typically such a kind of device is characterised by two concentric pipes which keep separated two main mass flow rates (green and red arrows in figure 1.14). The internal one is a standard circular pipe. The external pipe is fed by several tangentially oriented pipes placed at a specific distance from the main inlet which act like additional inlets (blue arrows). The peculiarity of these pipes is that they are tangentially oriented with respect to the axial direction. Hence, a mass flow rate coming from these pipes provides a rotating motion in the external pipe. In

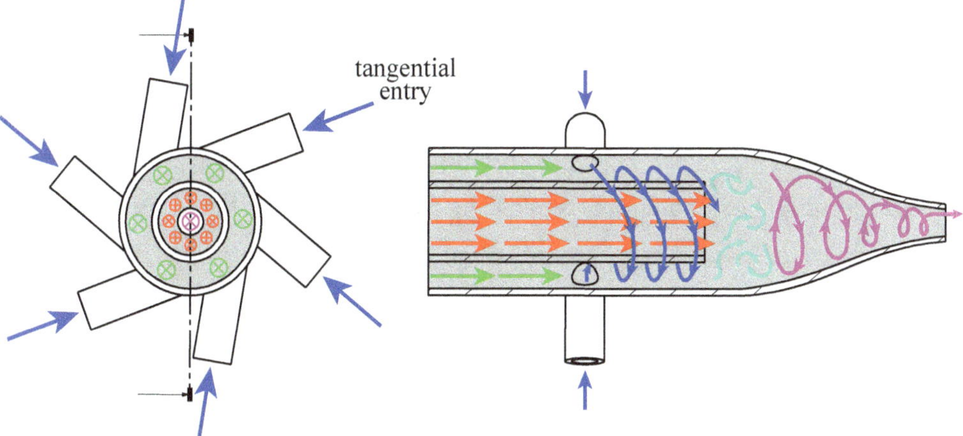

Figure 1.14: Schematic of an axial-plus-tangential entry swirl generator (Toh et al., 2010).

some cases, the mass flow rate, indicated with green arrows in figure 1.14, could not exist or, for reasons of design simplicity, it is equal to the one indicated by red arrows. Such a kind of device is largely applied when low swirl numbers are required. On the other hand, at high swirling rates the significant pressure drop makes preferable other design solutions.

The guided vane system is a valid alternative to generate swirling flows: it is simple to manufacture and requires easy maintenance. Swirl generators which involve this kind of design solution are characterised by several vanes oriented in order to deflect the flow direction (figure 1.15).

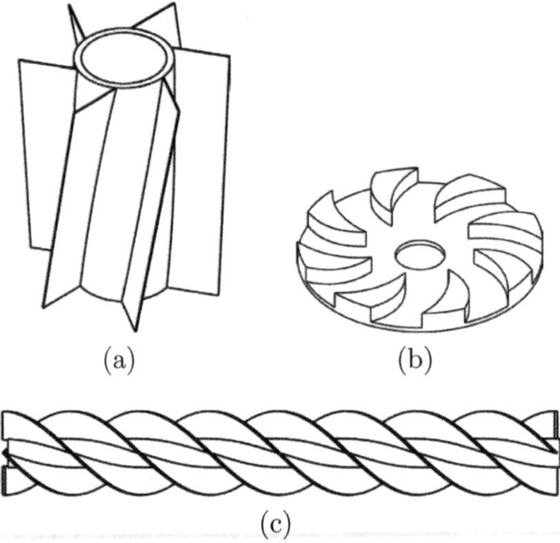

Figure 1.15: Different designs of a swirl generator with the application of the guided vane system.

In figure 1.15a the device is supposed to be placed in a pipe. Two paths appear to be defined: the external one, which generates the rotating motion, and the internal one, where the fluid straightly flows undisturbed. Such a kind of device is typically applied when there is full availability of space to let the flow develop. As a matter, once the two flows get together, it is fundamental leaving some characteristic lengths to let the flow reorganise and fully develop. One of the first practical realisations of this swirl generator design is reported by Chigier and Chervinsky (1967). The design parameters which tune the flow fields are the ratio between the internal and the external diameter, the vane pitch and the vane angle.

In some cases, the internal pipe can degenerate to a solid hub of small diameter. This means that the whole device become a multi-channel swirl generator (Ianiro et al., 2018).

The swirl generator depicted in figure 1.15b is supposed to work attached to a flat wall from the vanes' side (Jaafar et al., 2011). The design parameters are the vane pitch and the vane angle. In particular, the former affects the uniformity of the flow at the exit, the latter can be loosely interpreted. As a matter, a high swirl ratio can be obtained through a small vane degree but with an optimised custom vane shape (Susan-Resiga et al., 2008). A movable block swirl generator is the most flexible solution to rapidly vary the swirl ratio (Fudihara et al., 2003; Fudihara et al., 2007; Hohmann, 2011). In alternative designs the centred hole could also not exist. The whole device is very compact, so it is ideally suited for applications characterised by shortage of space.

As a final example of a swirl generator device, an object similar to which depicted in figure 1.15a, is represented in figure 1.15c. The device is characterised by several helices climbing over a straight solid hub. The characteristic parameters related to such a design are the pitch of helices and the ratio between hub and helix diameters. The first parameter affects the swirling degree of the flow, the second affects the flow rate (Bakirci and Bilen, 2007; Bilen et al., 2002). The degenerative version of such a device is characterised by the absence of the supporting hub resulting in a twisted tape swirl generator (Eiamsa-ard and Promvonge, 2005; Smithberg and Landis, 1964; Thianpong et al., 2009; Wen and Jang, 2003). The device is intended to work with high density, two-phase and reacting flows like in case of oil, sand or flames.

In case of design reported in figures 1.15a and 1.15c with the simplification of a solid hub and constant vane angle, the swirl number can be directly related to the vane angle.

$$S = \frac{2}{3} \left[\frac{1 - (D_h/D)^3}{1 - (D_h/D)^2} \right] \tan \varphi \tag{1.4}$$

where D_h is the hub diameter and φ is the constant vane angle (figure 1.16). Of course, in case of hubless devices the equation (1.4) results in

$$S = \frac{2}{3}\tan\varphi \qquad (1.5)$$

which means that vane angles of 15, 30, 45, 60, 70 and 80 degrees corresponds approximately to swirl number equal to 0.2, 0.4, 0.7, 1.2, 2.0 and 4.0, respectively.

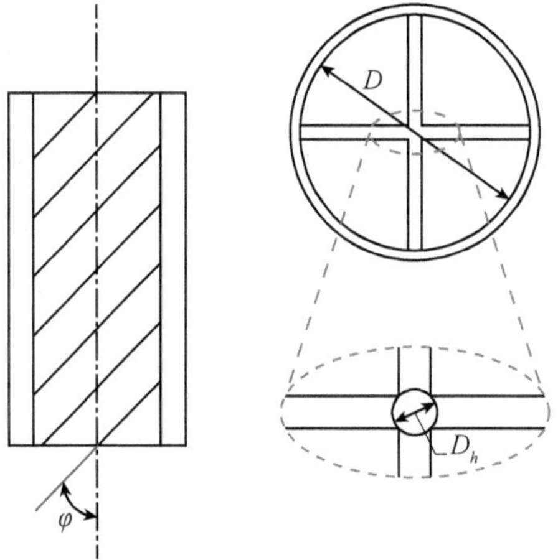

Figure 1.16: Characteristic parameters of a swirling nozzle with helical pattern (Ianiro and Cardone, 2012).

1.2.2 Free submerged swirling jets

The basics of theory about rotating flow fields was given by Görtler (1954) and Loitsyanskii (1953). Their aim was to extend the theory about turbulent jets to rotating flows. Both analyses were based on simplified equations of motion invoking the boundary-layer approximation. Loitsyanskii based his calculations including the radial pressure-gradient term. On the other hand, Görtler supposed to study a sufficient weak swirl degree assuming constant the pressure field throughout the jet. The aspects of decaying and spreading were studied by Steiger and Bloom (1962) who focused on the investigation of incompressible and compressible free jets with small, moderate, and large swirl degree. The tangential and axial velocity components and the stagnation enthalpy were assumed to have polynomial profiles in the radial direction. Then, the Von Kármán integral method was applied to the viscous layer but no comparisons with experiments were presented.

The analytical investigation continued with the work of Lee (1965) who obtained closed-form solutions for an axisymmetric turbulent swirling jet using similarity assumptions for the axial and the tangential velocities. The experimental data provided by Rose (1962) demonstrated to be in good agreement with the theory developed about the weak swirling case. Chigier and Chervinsky (1967) performed both theoretical and experimental studies of turbulent swirling jets emerging from a round orifice.

In the next years, investigations followed about new methods to generate swirling fluid motion. The investigations were conducted in a confined coannular configuration (Dolling and Gray, 1986). The interest on such a kind of flows increased in correspondence with the noticeable enhancements in combustion applications (Lilley, 1977; Syred and Beer, 1974). Turbo-machinery industry also took advantage of these advancements (Kerrebrock, 1977). The latest investigations included the study of the phenomenon of the vortex breakdown (or flow reversal) (Benjamin, 1962; Leibovich, 1978; Sarpkaya, 1971) and the vortex instability (Chanaud, 1965) in strong swirling flows.

As it can be noticed, the swirling flows investigation can be divided in five categories:

- swirling turbulent free jets emerging from an orifice and exhausting into a stationary or moving fluid;

- confined swirling flows in variable area ducts;

- swirling flows in turbo-machinery annuli;

- vortex control;

- leading-edge vortex breakdown over a high-angle-of-attack delta wing.

The present work will focus on the first framework.

The evolution of a swirling jet emerging from a nozzle into ambient fluid strongly depends on the method of swirl generation which basically affects the initial distribution of velocity (Farokhi et al., 1989; Gilchrist and Naughton, 2005). Hence, each experimental setup which involves the same swirl number could demonstrate, without any conflict with others' results, some differences in velocity profiles.

Figure 1.17 reports the radial distributions of the axial component of the velocity vector for a generic free swirling flow. As it can be noticed, there exists a different behaviour depending on the swirl number. From weak to moderate swirl ratios the velocity profiles appear to be much similar to Gaussian distributions. In addition, moving away from the jet axis, velocity profiles widen and decrease in maximum value. For strong and very strong swirl numbers a flow reversal phenomenon is measured

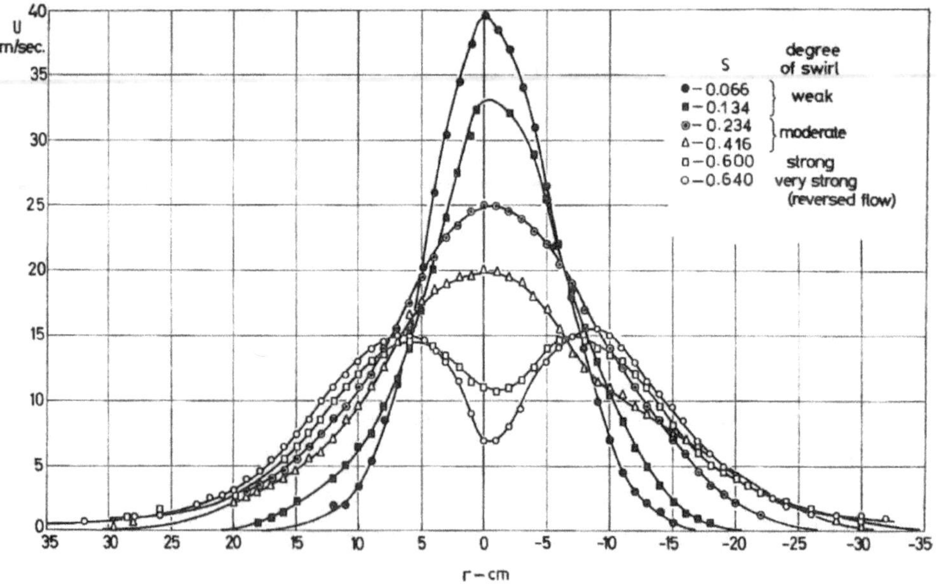

Figure 1.17: Radial distributions of axial velocity component at distance of $x/D = 4.1$ from the nozzle exit (Chigier and Chervinsky, 1967).

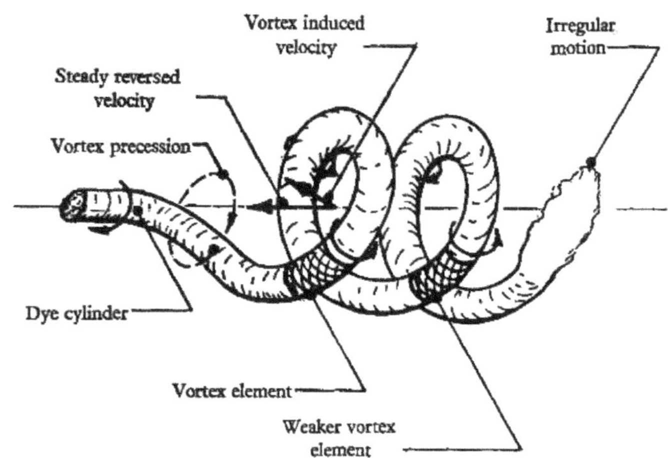

Figure 1.18: Schematic of periodic motion of the PVC and velocities induced (Chanaud, 1965).

in the region up to $3.5D$ from the orifice (Chang and Dhir, 1994).

The first three-dimensional investigation of a free swirling jet implied dye injection in water and air experiments performed on a swirl generator (Chanaud, 1965). Observations showed that, for swirl ratios lower than a critical value S_{cr} (Billant et al., 1998), a Kelvin-Helmholtz instability occurs in the axial shear layer caused by the high radial gradient of the azimuthal velocity in the nozzle centre (H. Liang and Maxworthy, 2005). Proceeding downstream, in the region of significant velocity deceleration, the swirling core takes the shape of a precessing helix (figure 1.18). This is a precessing vortex core (PVC) (Cala et al., 2006) which is the primary and the most powerful structure in a swirling jet flow. The dominant vorticity of the swirl flow is located in such co-rotating counter-winding helical structure. On the other hand, the outer shear layer located between the fluid at rest and the swirling flow appears in the form of tilted roll-up vortex rings (Markovich et al., 2014). Figure 1.19 show a schematic of the above description.

After a critical value of the ratio of rotational velocity to axial velocity ($S > S_{cr}$), free swirling jets are dominated by vortex breakdown and a central recirculation zone with a strong reverse flow. The flow experiences

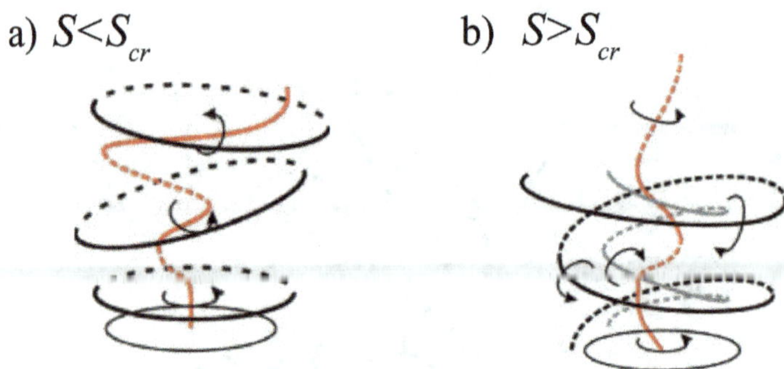

Figure 1.19: Schematics of the instantaneous structures of a) low- and b) high-swirl free jet (Markovich et al., 2014).

a regular temporally periodic motion of finite amplitude and the outer shear layer assumes a spiral shape called outer secondary vortex (OSV). In addition, the inner shear layer generates an inner secondary vortical (ISV) structure. The three identified vortical structures are schematically reported in figure 1.20. Such helical structures mainly differ in the energetic content.

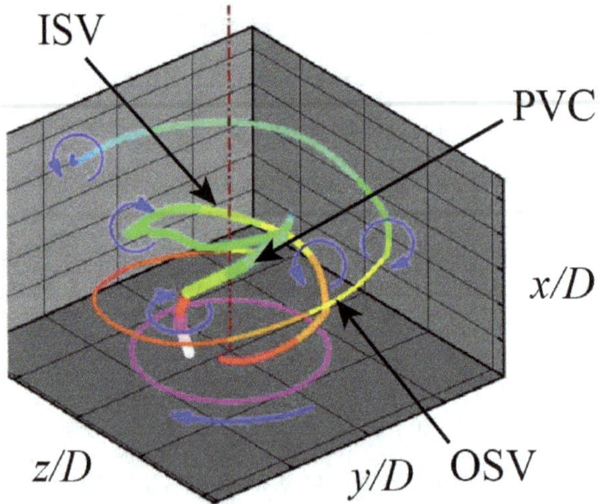

Figure 1.20: Schematic of a free turbulent swirling jet. The vortex signs are given by blue arrows. The purple circle represents the nozzle edge (Cala et al., 2006).

1.2.3   Impinging submerged swirling jets

One of the most significant characteristics of an impinging round jet is that the local heat flux is highly non-uniform (Viskanta, 1993). For several industrial applications, like electronic cooling and drying, this is not acceptable. Hence, the swirling jets could be a possible solution to achieve both high heat transfer and radial uniformity (Huang and El-Genk, 1998).

In figure 1.21 a schematic of the flow field of an impinging swirling jet is reported. The flow field is distinctly different from the free configuration.

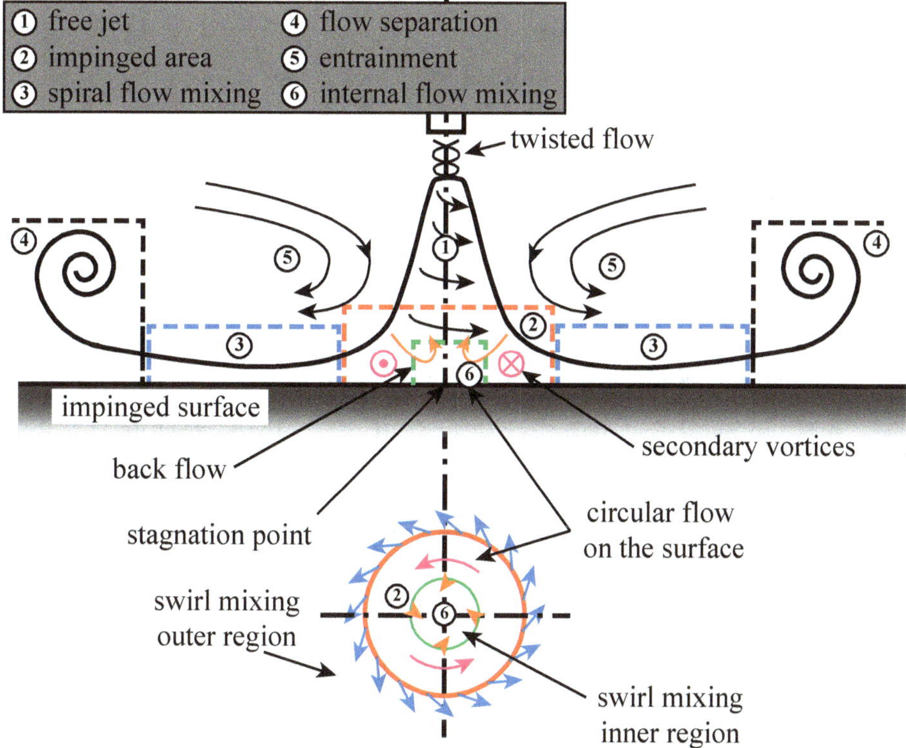

Figure 1.21: Schematic of the swirling impinging jet flow field (Huang and El-Genk, 1998).

Six different regions are noticeable: free jet region, impinging area, spiral flow mixing region, flow separation, entrainment region and internal flow mixing region.

The free jet region is comparable to what happens in a conventional impinging jet. The fluid proceeds toward the wall barely influenced by its presence. The rotating motion significantly affects the flow diameter spreading the jet and dragging fluid from the ambient.

The impinged area (marked with the dashed red line) is characterised by the impact of the jet with the wall. In this region, an annular rotating volume of fluids generates which further increases the mixing. With the approach of the fluid to the wall, the flow experiences a deflection along the wall. As a matter, the axial velocity component reduces to zero while the radial component of velocity rapidly increases. The azimuthal velocity component triggers the formation of a spiral flow which departs from the impact region and spreads developing in a wall jet. The jet finally separates at several diameters from the impingement.

The first investigations about swirling impinging jets are referred to Martin (1977). The conclusion of his work was that the presence of the circumferential velocity component did not produce significant changes on the heat or mass transfer from the wall. After a decade, Ward and Mahmood (1982) presented different conclusions demonstrating that swirl flow significantly reduces the heat or mass transfer from the wall. In order to solve such a contradiction, additional investigations were conducted about the instantaneous flow field (Abrantes and Azevedo, 2006; Alekseenko et al., 2007) and the heat transfer coefficient (Azevedo et al., 1997). The most striking difference between a conventional and a swirling jet was demonstrated to be the spreading in the radial direction caused by the presence of the swirl. This characteristic causes a decrease in the axial velocity in the center of the jet leading to the formation of zones of recirculating flow in the jet core. This gives rise to a second shear layer at the interface of the jet with the inner slow-moving-fluid region, where intense vortices are formed (figure 1.22a).

The presence of a recirculating zone at the stagnation region was confirmed with the combined application of velocity and temperature measuring techniques (Particle Image Velocimetry and Laser Induced Fluorescence) (Nozaki et al., 2003). Additional information were added by Senda et al. (2005) who applied Laser Doppler Velocimetry (LDV) and a

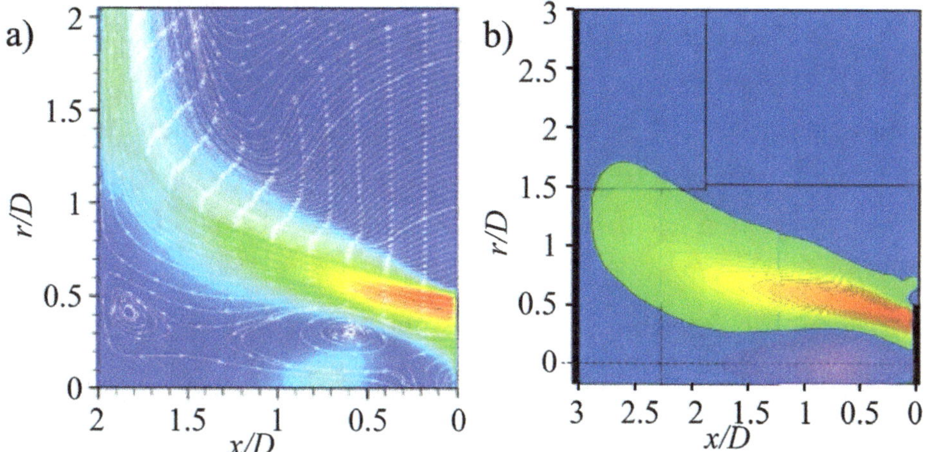

Figure 1.22: Quantitative measurements about the flow field between the nozzle and the impinging wall of an impinging swirling jet. a) Time averaged velocity magnitude contours (Abrantes and Azevedo, 2006). b) Normalised axial component of mean velocity contours (Alekseenko et al., 2007).

thermo-sensitive liquid crystal sheet to provide a detailed footprint of the thermal field. A local maximum of heat transfer was spotted at a non-zero radial position from the stagnation point. On the other hand, the heat transfer was demonstrated to be more uniformly distributed all over the impingement plate. Once verified the enhancement properties about heat and mass transfer of swirling impinging jets, many works followed (Brown et al., 2010; Hee et al., 2002; Kinsella et al., 2008; Nuntadusit et al., 2010).

In order to provide additional information to make the interpretation of the velocity in the third dimensions easier, a recent work of Ahmed et al. (2015) gave the first quantitative measurements about the pressure distribution in the stagnation and wall jet regions for relatively high swirl numbers.

1.3   Mixing and transport of passive scalars

Entrainment phenomena are always present in both isothermal free and impinging jet configurations, swirling or not. When a jet emerges from a nozzle, a friction occurs between the surrounding fluid at rest and the forced motion.

As a matter, entrainment acts a main role in the mixing performances of a jet. Hence, in configurations in which the jet is at different temperature with respect to the ambient in which exhausts, it is interesting to investigate the mixing properties provided by the jet under analysis (Dahm and Dimotakis, 1987; Strang and Fernando, 2001).

The mixing performances of a jet have fundamental importance in impinging configurations. The interaction of the main flow with the surrounding fluid occurs in multiple regions and in a more complex way because of additional phenomena involved in the impact of the jet. In impinging jet configurations, entrainment phenomenon can significantly affect the heat transfer performances acting as driver of new fluid toward the jet axis. Attracted by heat transfer enhancement, several investigations focused on flow fields large-structure characterisation always followed by heat transfer analyses (Afroz and Sharif, 2018; Amini et al., 2015; Fénot et al., 2015; Li et al., 2018). Such investigations suggest a strict correlation between entrainment and mixing brought by large-scale vortical structures involved in the flow field.

The most immediate way to investigate the mixing performances of an impinging jet is to take advantage of transport of passive scalars provided by the flow field under analysis. Hence, implying a jet at different temperature from the ambient in which exhausts, the different radiative content of two fluids acts as a tracer which has a diffusive behaviour but has no influence on the dynamics of the flow field (Warhaft, 2000). As a matter, the information provided by simultaneous acquisition of unsteady velocity and temperature fluctuations fields can be applied to correlate the main features related to the two measured quantities (Antoranz et al.,

2018; Haustein et al., 2012; Yamada and Nakamura, 2016).

1.4 Motivation of the work

The interaction of a swirling jet with a flat wall is of main interest for several research topics because of the entrainment and mixing enhanced properties. Despite that, literature is currently lacking of a detailed three-dimensional description of the flow field in proximity of the wall and its instantaneous influence on the thermal field. For this reason, the investigation of both the unsteady three-dimensional velocity and thermal fields in proximity of the wall of an impinging swirling hot jet exhausting in a cold ambient is attractive. Hence, the present work aims to accomplish the following goals:

- design of a custom swirl generator with variable vane angle;

- design of an experimental apparatus which allows to simultaneously measure the three-dimensional unsteady flow field dynamics and the thermal fluctuations in proximity of the wall;

- characterisation of the swirl numbers related to the custom swirl generators designed;

- three-dimensional description of the flow field of a swirling jet in proximity of the nozzle;

- comparison of several swirl generators mixing performances examining the unsteady and simultaneous impinging three-dimensional flow field and wall temperature;

- application of modal decomposition methods to extract the main vortical structures involved in a swirling jet;

- estimation of the correlation between impinging flow field and wall temperature.

In the following, we first present a non-exhaustive literature review of the measurement techniques applied to perform the experimental tests: time-resolved Tomographic Particle Image Velocimetry (TR Tomo-PIV) and Infrared (IR) thermography. The working principles and the peculiarities of each technique are highlighted in order to give to the reader the basic elements to critically understand the challenge of the simultaneous acquisition of both velocity and thermal field of an impinging swirling jet. Such measurements are performed with the experimental setup described in chapter 3. The design characteristics of the employed swirl generator, the details of the images processing and the limits of the measurement are described in detail. Then, the two reference systems adopted for the experimental data analysis, one in proximity of the nozzle and another one in proximity of the wall, are defined.

In chapter 4, the flow field in proximity of the nozzle of the swirl generator is defined with the description of the radial velocity profiles and of the three-dimensional instantaneous flow field. Then, with the application of Proper Orthogonal Decomposition analysis (Berkooz et al., 1993), the main features of a free swirling jet are presented.

The characteristics of the flow field of an impinging swirling jet are investigated for five swirl numbers. Separate analyses for the velocity field and for the temperature field are presented. Then, the characteristics of the simultaneous unsteady velocity and temperature flow field are described. As final remark, the correlation between the two measured quantities is presented with the description of the outcome of two different correlation techniques.

2 | Measurement techniques

In order to successfully measure the three-dimensional velocity and temperature distributions in proximity of the wall, two techniques are simultaneously applied: Tomographic Particle Image Velocimetry (PIV) and Infrared thermography. In the following, the basic principles of the two techniques will be described.

The first mentioned technique is notorious to have been largely employed to provide three-dimensional time-resolved (4D measurements) data. Hence, with this technique, it is possible to investigate the time-dependent three-dimensional organisation of turbulent coherent structures.

The application of the second technique allows to detect the electromagnetic energy radiated by an object and to convert it into a sequence of instantaneous 2D temperature maps. As a matter, an IR camera offers fully 2D non-intrusive high sensitive and low response time measurements of the accessible field of view.

2.1 Tomographic Particle Image Velocimetry technique

The major drawback of most of quantitative techniques is placing a sensor, more or less sophisticated, in proximity of or within the flow to be measured. In both cases the flow, especially in subsonic regimes, inevitably feels the presence of the sensor modifying its topology.

Driven by the need to investigate in detail the flow field without affecting it in any way, together with the technical progress about high-energy lights (lasers) and fast imaging technology, after about a century from the first PIV-oriented measurements (Prandtl, 1904), the formal bases of the techniques were posed (Elsinga et al., 2006).

2.1.1   Fundamentals of technique

Tomographic PIV (T-PIV) technique is a completely non-intrusive and whole-field technique which allows the measurement of velocity vector in a three-dimensional volume exploiting the light scattered by small particles uniformly disseminated in the fluid which are assumed to follow the fluid flow with high accuracy. A stroboscopic high-energy light source, typically a laser light, is used to illuminate the seeding particles at two consecutive instants separated by a small time delay. The light scattered by these particles is then collected by multiple high-speed cameras.

A typical T-PIV experimental setup is schematically represented in figure 2.1 where also the processing steps are reported. The images acquired are then reconstructed to form three-dimensional intensity distributions. The field of displacements is then evaluated via statistical techniques.

Indeed, the above described process includes some additional steps which play a fundamental role for the accuracy of the process. First of all, in order to obtain an accurate 3D reconstruction, a calibration of the camera system is needed. This includes two steps: a calibration procedure which includes a physical target and a refined calibration from particle images called *self-calibration* (Wieneke, 2008).

2.1.2   Volume illumination and optical system

A PIV optical system includes two main components: a light source and several digital cameras. In most PIV applications, the illumination of the measurement volume is typically performed using a pulsed laser. Such

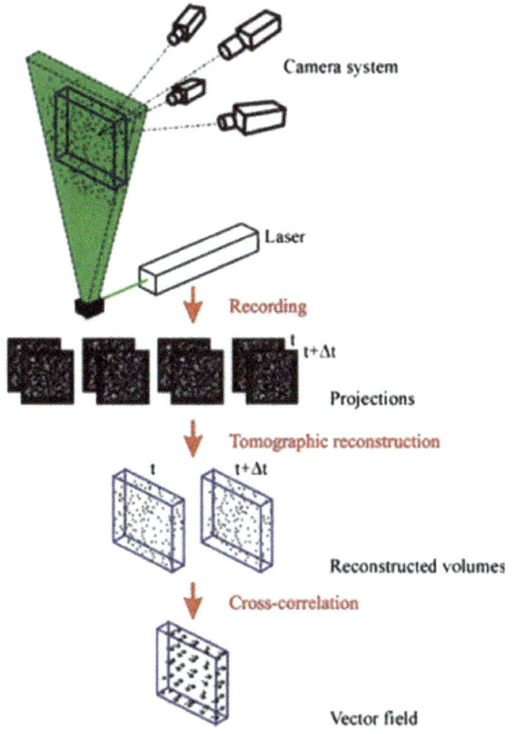

Figure 2.1: Schematic of T-PIV setup and processing procedure (Elsinga et al., 2006).

a device provides high energy (10-400mJ/pulse) and short duration light emission (6-100ns). The first characteristic is fundamental to achieve sharp distinction between the particles which scatter light and the background which is typically close, in value, to the camera image noise. The second is necessary to avoid blurred particles patterns. High energy pulses can be achieved only from low to moderate repetition rates (up to 15Hz). For low-speed applications the most appropriate devices are the Nd:YAG lasers. For high-speed applications lower pulse energy (15mJ to 30mJ) lasers are used. These devices, typically Nd:YLF lasers, have the operating frequency ranging between 10Hz to 10kHz. The wavelength of the emitted light is 532nm for the Nd:YAG lasers and 526nm for the Nd:YLF ones.

Common PIV cameras can be distinguished for the sensor mounted on them: CCD (charge-coupled devices) or CMOS (complementary metal-oxide-semiconductor) sensors. The first kind of sensors is typically slower and more expensive but offer higher dynamic range and higher resolution. On the other hand, CMOS sensors allow for repetition rates up to few thousands per second. The more recent sCMOS (scientific CMOS) technology combines the advantages of modern CCD and CMOS sensors. The typical sensor sizes vary between 1 to 16 millions of pixels with pixel size ranging between 5µm to 25µm.

In order to achieve high accuracy in measurements, particles require to be in focus throughout the illuminated region and to have images on the sensor plane with an appropriate size (in pixel units). To achieve such conditions, cameras have to be equipped with lenses.

The diameter of particles on the camera sensor d_τ can be spilt in two contributions: the particle diameter related to geometric effects (d_{geom}) and which related to diffraction effects (d_{diff}). The first is related to the optical magnification and the second is related to lens aperture and light wavelength. The particle diameter for a specific setup can be easily *a priori* evaluated with the application of the following (Adrian and Yao, 1985):

$$d_\tau = \sqrt{d_{\mathrm{geom}}^2 + d_{\mathrm{diff}}^2} = \sqrt{(Md_p)^2 + [2.44\lambda f_\#(1 + M)^2]^2} \qquad (2.1)$$

where d_p is the physical particle diameter, M is the magnification factor equal to d_{geom}/d_p, λ is the wavelength of illumination light and $f_\#$ is the f-number equal to the ratio of lens focal length to aperture diameter. Typically, d_τ is expressed in pixel units as $d_\tau^* = d_\tau/d_{\mathrm{pix}}$, with d_{pix} being the pixel size in physical units. As a rule of thumb, d_τ should be greater than 2 and smaller than 5. The first constraint should be respected to avoid "peak locking" issues, the second one should be verified to keep high the accuracy of the computation.

The condition of focused particles is accomplished when all illuminated particles included in the field of view do not encounter a significant blurring. The extension of the in-focus region is referred to as "depth of field" and it is defined as (Raffel et al., 2018):

$$\Delta_{\text{field}} = 2f_{\#}d_{\text{diff}}\left(\frac{M+1}{M}\right) = 4.88f_{\#}^2\lambda\left(\frac{M+1}{M}\right)^2. \tag{2.2}$$

In case of volumetric measurements, the cameras are typically tilted with respect to the midplane of measurement and the required depth of field can be large. The first difficulty can be overcome with the application of specific lens adapters those make parallel the plane of focus to the image plane (Scheimpflug condition). The second issue can be solved increasing the $f_{\#}$ which also causes the reduction of light acquired by the sensor.

2.1.3   Optical system calibration

In order to convert the measured displacements from pixel (or voxel, the equivalent of pixel in 3D space) in physical units, a calibration procedure is needed. In particular, for tomographic PIV measurements this step is necessary for the reconstruction procedure. The modern camera calibration procedure consists of two main steps: a calibration which includes a physical target and a refined calibration which exploits particle images.

Physical calibration

The *physical calibration* is based on the registration of a calibration target covered by markers whose physical positions are precisely known (figure 2.2). The calibration target is typically mounted on a translation stage allowing images to be acquired with various depth positions. The position of each marker is determined on the images by template matching algorithms. The result of this mark finding procedure is the correspondence of image with world positions. Such a correspondence is used to define an imaging model for the optical system. Typically, the pinhole camera model (Tsai, 1987) is applied because of its ability to describe the camera in terms of physical parameters. On the other hand, such a camera model is limited in its ability to describe lens distortions. Alternatively, a poly-

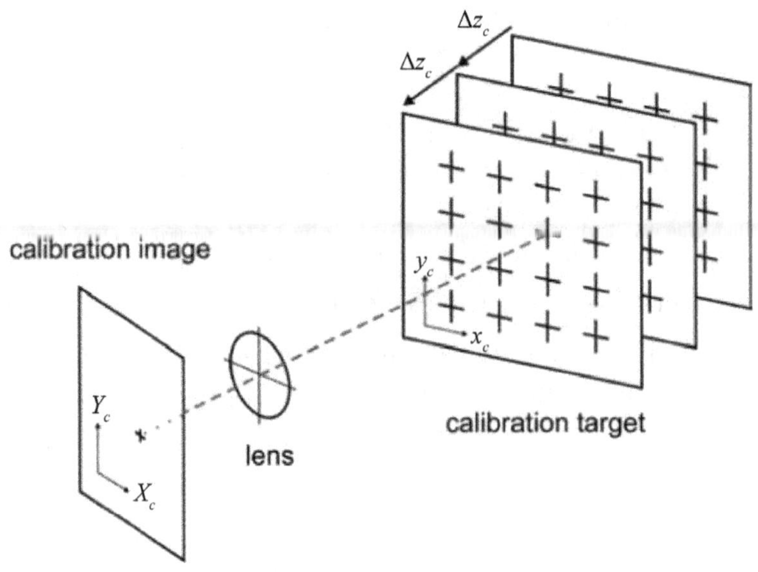

Figure 2.2: Schematic of physical calibration procedure with the acquisition of multiple calibration target positions (Elsinga, 2008).

nomial model for higher-order distortions to be captured can be defined. A least-squares fitting method is applied between the marker positions in world reference system (x_c, y_c, z_c) and those on images recorded (X_c, Y_c). The fit returns a set of polynomial coefficients rather than camera parameters. Typically, for thin volumes the method proposed by Soloff et al. (1997) is applied. This method involves a third-order in-plane and second-order out-of-plane polynomial fitting. For thick volumes independent third-order in-plane polynomials are used for each depth plane, called *mapping plane*. The intermediate depth positions are determined by linear interpolation of two adjacent mapping planes. The accuracy of a model can be estimated by the root-mean-square (RMS) residual error between the measured and modelled camera positions. Typically the accuracy achieved with such a procedure is of 0.5 pixels.

Self-calibration procedure

The accuracy of optical system calibration presented above can be further improved using a self-calibration procedure using the particle images. Errors from physical calibration could arise from target imperfections, accuracy of marker identification, incorrect positioning of target in the world coordinate system. The volume self-calibration proposed by Wieneke (2008) solves these issues through the use of particle triangulation. The triangulation procedure is schematically reported in figure 2.3.

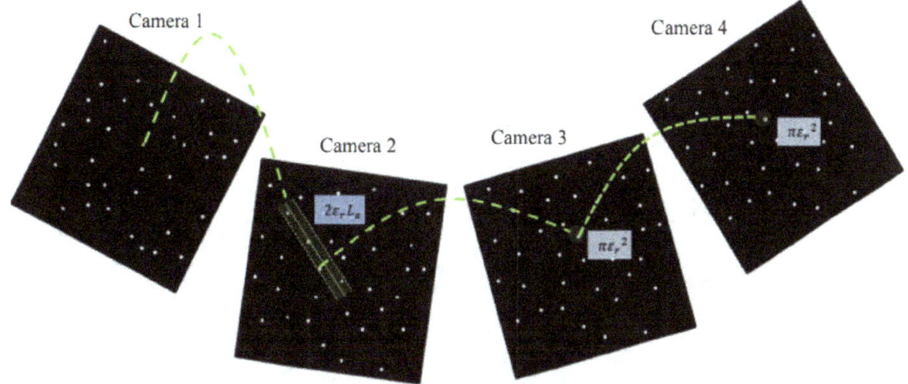

Figure 2.3: Particle matching search procedure for four cameras; ε_r is the search radius, L_z is the length of the epipolar line (Discetti and Astarita, 2014).

First, a set of particles are identified via intensity thresholding method and 2D peak detection on the first camera. Then, the line of sight of each particle position is computed and back-projected onto the second camera. The back-projected line of sight (LOS), intersecting the illuminated volume, is an epipolar line. Established a search radius ε_r, all particles located within the region are flagged as potential matching candidates. The procedure continues on the other cameras. It is worth noting that after two cameras, the epipolar line is an epipolar point. This means that the search radius highlights a circular region. If a match is found in all camera images, the particle is considered a true match and a least-squares triangulation estimates the particle position in world coordinates. The back-projection of such a position to all cameras and comparison to the particle images provides the error map for each camera at the world coordinates of the particle. The disparities are collected in a binning procedure, each modelled as a Gaussian blob integrated onto a disparity map. As additional disparities are determined, a peak forms indicating the most likely disparity. To improve the disparity peak detectability, the procedure is repeated for a number of snapshots and the disparity maps for the individual snapshots are binned. The volume self-calibration is capable of correcting calibration errors in excess of 10 pixels and reducing them to below 0.1 or even 0.01 pixels (Wieneke, 2008).

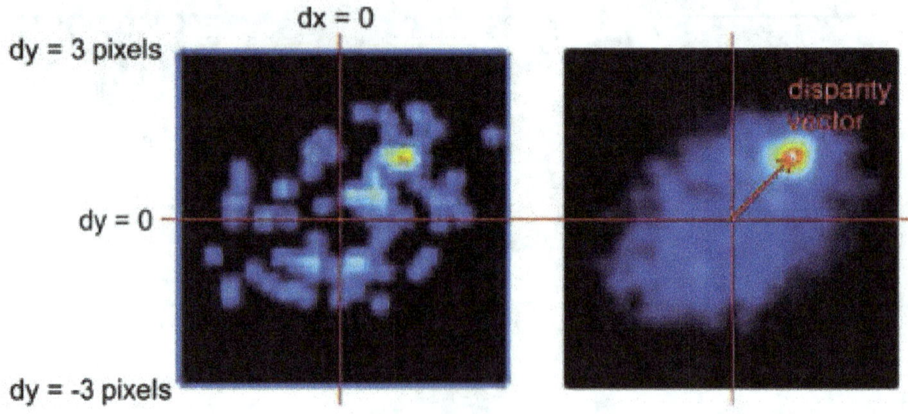

Figure 2.4: Disparity maps generated from self-calibration procedure for a single image (left) and for 16 images (right) (Wieneke, 2008).

2.1.4 Tomographic reconstruction

The tomographic reconstruction procedure aims to reconstruct the three-dimensional intensity distribution E starting from its projection on the cameras denoted with I. Typically, the optical medium in which experiments are performed is transparent, I is the result of the integration of the intensity field E along each pixel LOS. Such a relation can be described by a linear system of equation

$$\sum_{j=1}^{m} \omega_{ij}^{(c)} E_j = I_i^{(c)} \quad \text{with} \quad i = 1, \ldots, n, \quad c = 1, \ldots, N \tag{2.3}$$

where n is the number of pixels, m is the number of voxels, i is the index over pixels of the c camera and j is the index over the voxels of the intensity fields. The coefficients $\omega_{ij}^{(c)}$ are the weighting functions which specify the contribution of each jth voxel to the ith pixel. Such weights, which are estimated by the volumetric fraction of voxel j intersected by the LOS of pixel i, are physically estimated by the intersection of a pyramid, representing the pixel LOS, and a cubic voxel. To reduce the computational costs the LOSs are assumed to be cylindrical and the voxels spheres.

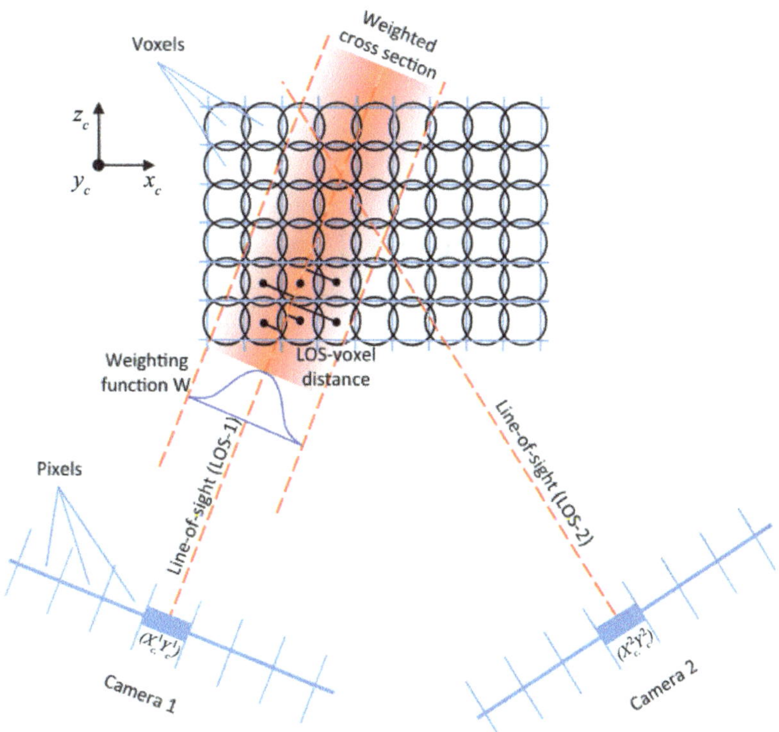

Figure 2.5: Schematic of voxels discretisation and imaging model adopted in tomographic reconstruction. The weighting function for the pixel belonging to the camera 1 has a Gaussian distribution and only the voxels falling in this region contribute to the light intensity of the pixel (Scarano, 2012).

The weights in equation (2.3) forms a non-square and non-symmetric matrix. This precludes to solve the $\underline{\omega}\,\underline{E} = \underline{I}$ system with the computation of the inverse matrix.

The algebraic schemes applied to solve such a problem can be distinguished in two classes: additive and multiplicative. The standard methods belonging to these two classes are ART, algebraic reconstruction technique (Gordon et al., 1970), and MART, multiplicative algebraic reconstruction technique (Herman and Lent, 1976). In the following the iteratively applied equations belonging to the two reconstruction techniques are reported:

$$\text{ART:} \qquad E_j^{(k+1)} = E_j^{(k)} + \bar{\mu}\frac{I_i^{(c)} - \sum_j \omega_{ij}^{(c)} E_j^{(k)}}{\sum_j \omega_{ij}^{(c)}}\omega_{ij}^{(c)} \qquad (2.4)$$

$$\text{MART:} \qquad E_j^{(k+1)} = E_j^{(k)} \left(\frac{I_i^{(c)}}{\sum_j \omega_{ij}^{(c)} E_j^{(k)}}\right)^{\bar{\mu}\omega_{ij}^{(c)}} \qquad (2.5)$$

where $\bar{\mu}$ is a relaxation parameter useful to stability of the method.

A quantitative comparison of reconstruction quality provided by different methods is provided by Elsinga et al. (2006). In the case where a reference volume E_{ref} is known, like in synthetic tests, the reconstruction quality factor Q_f is defined as the normalised cross-correlation coefficient between a sample volume E and E_{ref}:

$$Q_f = \frac{\sum_j^N E_j \cdot E_{j,\text{ref}}}{\sqrt{\sum_j^N E_j^2 \cdot \sum_j^N E_{j,\text{ref}}^2}}. \qquad (2.6)$$

In case of experimental data, Q_f is estimated as the residual error between the back-projection of the reconstructed volume on the image plane and the acquired images. The Q_f value is bounded between 0 and 1 indicating that 1 is an ideal reconstruction procedure.

2.1.5 Advanced reconstruction techniques

In order to increase the reconstruction quality several approaches of the reconstruction problem have been tested. The main source of error during the reconstruction step is caused by *ghost particles*. Ghost particles are regions of non-zero light intensity formed at the intersection of LOSs where no actual particle is indeed located in the physical space. Figure 2.6 schematically shows the detrimental effect of the ghost particle presence in a 2-particle-2-camera problem.

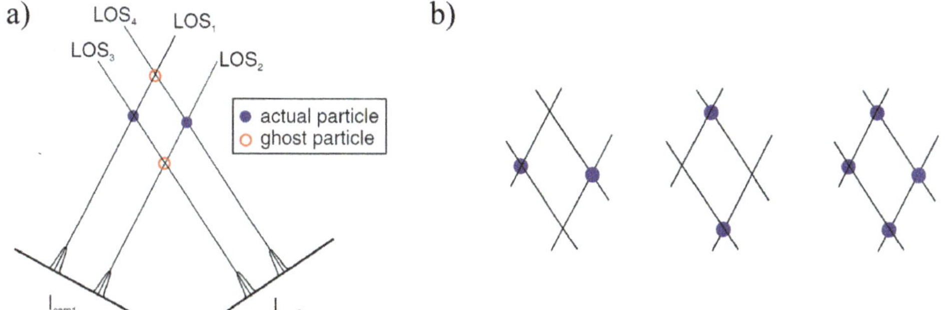

Figure 2.6: a) Schematic of the formation of ghost particles in tomographic reconstruction in 2-particle-2-camera problem; b) possible reconstruction solution to the 2-particle-2-camera problem reported on the left (Elsinga et al., 2011).

Increasing the number of cameras can help but it is not a definitive solution. In case of N-camera system, the estimation of the ratio of the number of ghost particles N_g to the number of true particles N_p is given by (Discetti, 2013):

$$\frac{N_g}{N_p} = ppp^{N-1} \left(\frac{\pi d_\tau^*}{4}\right)^{N-2} d_\tau^* \frac{M\Delta_{\text{field}}}{d_{\text{pix}}} = \frac{4}{\pi d_\tau} N_S^{N-1} M\Delta_{\text{field}} \qquad (2.7)$$

where *ppp* is the particle per pixel area which can be interpreted as the particle image density, N_S is the source density given by the product of *ppp* and the average particle area in pixels. The optimal value of N_S is around 0.2 (the total area occupied by particles in the recording is the 20% of the total imaging area), which corresponds to *ppp* $\approx$ 0.05 (Scarano, 2012). Beyond this threshold, the detrimental effect of the ghost particles leads to unacceptable levels of uncertainty in velocity field estimation.

Alternative strategies to ART/MART aim to reduce the number of ghost particles exploiting the coherence of a particle field between consecutive snapshots to present an improved initial guess to the MART procedure. For this reason, such a kind of approaches is suitable for time-resolved measurements. Enhancements are possible introducing the time information like in sequential motion tracking-enhanced (MTE) approach (Lynch and Scarano, 2015; Novara et al., 2010).

A different approach, called three-dimensional Particle Tracking Velocimetry (3D-PTV), considers discrete particles coupled with a particle imaging model instead of the classic voxel-based reconstruction logic. Such an approach is based on the particle detection and an iterative triangulation procedure (Maas et al., 1993). By the pairing of the particles between two consecutive snapshots it is possible to determine the velocity of tracer particles. Improvements in such a method have been provided by the iterative particle reconstruction (IPR) technique (Wieneke, 2012) and the 'Shake-the-Box' (STB) approach proposed by Schanz et al. (2016) and Schröder et al. (2015).

The IPR technique assumes that a single particle is not defined only by its position in the physical space but also by its diameter and intensity. Iterations are needed for optimizing all these properties. This is accomplished projecting the 3D particle distribution onto the cameras and comparing the back-projections with the original images. Hence, the properties involved in the optimisation process are refined minimising the norm of the residual images, defined by the difference between the back-projection and the original image. In such an iterative procedure, particles those show intensity or diameter below certain thresholds are identified as ghost particles and removed.

The IPR technique combined with time-resolved 3D-PTV allows for operating at higher seeding densities. The STB algorithm, which can be considered as the motion-tracking enhancement variant of IPR, overcomes the typical limitations in particle image density for particle position based three-dimensional methods introducing the temporal information into the IPR processing of each time-step predicting a particle cloud for each consecutive time-step (time-resolved STB). The particles paths are extrapolated in time and such positions are refined by 'shaking' the 3D particle distribution, then new particles are introduced and refined by IPR procedure.

2.1.6 Velocity estimation

The three-dimensional velocity field is obtained by the displacement evaluation of particle tracers between two or more consecutive snapshots. The displacement all over the tomographic PIV volume is obtained by dividing the reconstructed volume into a number of sub-volumes (interrogation volumes) and for each of them the three-dimensional cross-correlation function R is applied:

$$R(\underline{s}) = \frac{\sum_{\underline{x}}(\Omega_n(\underline{x}) - \xi_n) \cdot (\Omega_{n+1}(\underline{x} + \underline{s}) - \xi_{n+1}(\underline{s}))}{\sqrt{\sum_{\underline{x}}(\Omega_n(\underline{x}) - \xi_n)^2 \cdot \sum_{\underline{x}}(\Omega_{n+1}(\underline{x} + \underline{s}) - \xi_{n+1}(\underline{s}))^2}} \tag{2.8}$$

where $\underline{x}$ is the volume index, $\underline{s}$ is the vector of shifts, Ω_n and Ω_{n+1} are the interrogation volumes for two different times t and $t + \Delta t$, ξ_n and ξ_{n+1} are the mean values of Ω_n and Ω_{n+1} respectively. The function $R(\underline{s})$ ranges between -1 and 1. The peak location of the three-dimensional cross-correlation map is the estimated average displacement within the interrogation volume.

The maximum value of the correlation peak is negatively influenced by two effects: the out-of-window displacement of particles and the presence of velocity gradients within the interrogation window. The first effect requires large interrogation windows, the second can be easily avoided with smaller windows. As a matter, the two effects act simultaneously, thus to avoid both of them, a multi-grid iterative window deformation (MGIWD) is generally applied (Scarano, 2001). This approach consists in determining a predictor of the displacement field on a coarse grid and accordingly deforming the interrogation windows.

2.2   Infrared thermography

Infrared thermography is a measurement technique that enables to obtain non-intrusive measurements of surface temperatures by detecting the energy radiated in the IR band from any body at a temperature above absolute zero (Astarita and Carlomagno, 2012).

The temperature measurement in thermo-fluid-dynamics requires a temperature transducer. Most of the quantitative techniques involve zero dimensional sensors which are able to measure temperature at a single point. This is a significant limitation when the temperature field exhibits high spatial and temporal variations. There are several advantages in the application of an IR camera as a temperature transducer. Modern transducers have high sensitivity (down to 10mK) and low response time (down to 20µm). Hence, an Infrared camera constitutes a truly two-dimensional transducer, allowing accurate temperature measurements at high spatial and temporal resolution.

2.2.1   Principles of radiation

The heat transfer by radiation is an energy transport mechanism that occurs in the form of electromagnetic waves (Howell et al., 2016).

Suppose that a certain amount of energy is travelling in vacuum via radiation mode. The energy can be partially absorbed and reflected by a body or even pass through it. A conservation condition has to be verified:

$$\alpha_r + \chi_r + \tau_r = 1 \tag{2.9}$$

where α_r is the radiation fraction absorbed by the body, χ_r is the fraction of energy reflected by the body and τ_r is the fraction of total energy which passes through the body. Such contributions are respectively named absorptivity, reflectivity and transmissivity coefficients of the body under consideration. They depend on radiation wavelength and propagation direction.

The body which emits the greatest amount of energy at a given temperature is called *black body*. The law that describes the energy flux (energy rate per unit body area) per wavelength (spectral hemispherical emissive power) $P_b(\lambda)[\text{W/m}^3]$ emitted by a black body in the hemisphere outside its surface, is the Planck's law of radiation:

$$P_b(\lambda) = \frac{K_1}{\lambda^5 \left(e^{K_2/\lambda T_b} - 1\right)} \tag{2.10}$$

where λ is the radiation wavelength [m], T_b is the absolute black body temperature [K], K_1 and K_2 are the first and the second universal constants respectively equal to $3.7418 \cdot 10^{-16} \text{Wm}^2$ and $1.4388 \cdot 10^{-2} mK$. Equation (2.10) is graphically represented in figure 2.7. The curves reported tend to zero for both $\lambda \to 0$ and $\lambda \to \infty$.

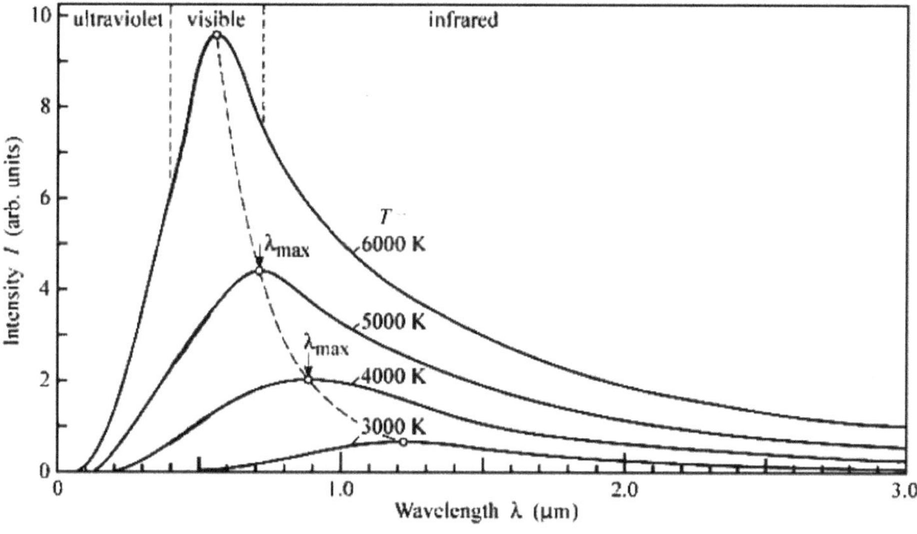

Figure 2.7: Black body radiation curve according to the Planck's law.

The entire electromagnetic spectrum can be divided into a number of wavelength intervals called *spectral bands*. The spectrum extends from very small wavelength values ($\lambda \to 0$) to extremely large ones ($\lambda \to \infty$). The Infrared range is included from few micrometers to some millime-

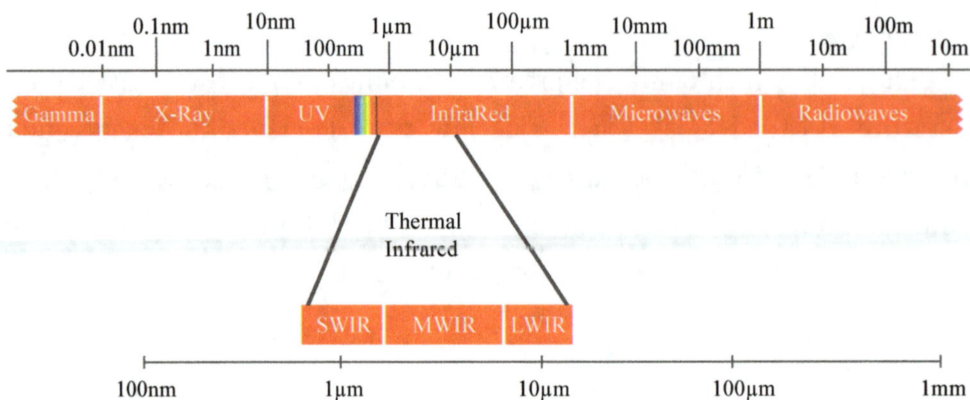

Figure 2.8: Electromagnetic spectrum with particular focus to thermal Infrared range.

tres (figure 2.8). The Infrared range can be further divided into four different bands: near- or small-wave IR (NWIR/SWIR) (0.75-3μm), mid-wave IR (MWIR; 3-6μm), long-wave IR (LWIR; 6-15μm) and extreme IR (>15μm). Most of the currently employed IR camera detectors are sensitive to MWIR or to LWIR spectral bands.

By deriving and integrating the Planck's law with respect to λ leads respectively to two other laws: Wien's displacement law and Stefan-Boltzmann's law.

The Wien's displacement law states that the wavelength λ^* at which the black body emits its maximum emissive power is a function solely of the black body temperature T_b:

$$\lambda^* T_b = 2897.8 \mu\text{mK}. \tag{2.11}$$

According to this law the location of maximum value of P_b moves toward shorter wavelengths as the body temperature increases.

The integration of the Planck's law provides a functional dependence between the emissive power E_b integrated on the whole spectrum of wavelength and the fourth power of black body temperature:

$$E_b = \beta T^4 \tag{2.12}$$

where β is the Stefan-Boltzmann's constant equal to $5.6704 \cdot 10^{-8} \mathrm{W/m^2 K^4}$.

Real objects never satisfy the laws described above because they are often a good approximation of a black body only in a specific spectral band. Hence, a real object generally emits only a fraction $P_r(\lambda)$ of the radiation emitted by a black body $P_b(\lambda)$ at the same temperature and at the same wavelength. Equation (2.10) can thus be rewritten as following:

$$P_r(\lambda) = \epsilon(\lambda) \frac{K_1}{\lambda^5 \left(e^{K_2/\lambda T_b} - 1 \right)} \tag{2.13}$$

where $\epsilon(\lambda)$ is the spectral emissivity coefficient defined as $\epsilon(\lambda) = P_r(\lambda)/P_b(\lambda)$.

2.2.2 Infrared cameras: hardware and performances

The core of the IR technique application is the radiation detector. Such devices are typically divided into two classes differing one from each other for the working principle: thermal detectors and photon detectors.

The working principle of the first class is that the incident radiation changes one of the electrical properties of the detector due to a temperature modification of the detector itself. In the second, the incident photons directly interact with the electrons of the detector material. In the detectors belonging to the former class the electrical response is proportional to the absorbed energy, for those belonging to the latter class the electrical response is dependent to the number of absorbed photons that interact with the sensor.

Figure 2.9 reports a schematic of the application of IR technique. The electromagnetic radiation waves originated from a specific scenario are collected in an IR-transparent objective. The radiation causes the excitation of pixels of the thermal sensor which is sensible to a selected IR band. A modern Infrared temperature detector for IR cameras is formed by a two-dimensional array of sensible elements that is normally called Focal Plane Array (FPA) detector. The electronic signal generated

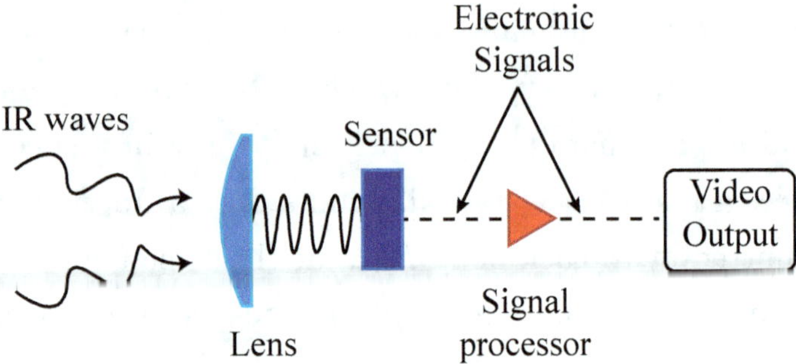

Figure 2.9: Schematic of the working principle of Infrared thermography technique.

is processed and converted in video output.

As it can be noticed, the main difference with respect to an optical system is that the material that constitutes the optical components has to be transparent in the investigated Infrared band. As a matter, most of the optical lenses, transparent in the visible band, normally appear opaque in IR bands. Typical optical materials used for general Infrared applications are: sapphire, calcium fluoride, magnesium fluoride for MWIR, while germanium and zinc derivatives for MWIR and LWIR bands.

The performance of an Infrared camera is typically evaluated in terms of four different characteristics: thermal sensitivity, scan speed, sensor size and dynamic range.

Thermal sensitivity is defined in terms of mean noise equivalent temperature difference (NETD), which is the time standard deviation averaged over all pixels of a black-body scene. Modern FPA are able to detect temperature differences of less than 20mK.

The scan speed represents the rate at which a frame is updated and is expressed in terms of frame rate (Hz). Modern FPA systems are able of hundreds of Hz. Typically the scanning rate can be increased cropping the sensor or reducing the integration time (IT), which is basically the duration in µs of sensor exposition to the electromagnetic radiation.

The sensor size (px × px) refers to the number of the sensor pixels.

The last parameter which define the performances of an IR camera is the dynamic range which is expressed in terms of the number of digital intensity levels the thermal image is composed of.

2

3 | Experimental setup

The three-dimensional flow field of an impinging swirling jet and its thermal footprint on the impingement target are measured by applying Tomographic Particle Image Velocimetry and Infrared thermography techniques. The experimental tests are carried out in an experimental apparatus whose main components will be described in this section. In addition, details will be given on the experimental working conditions and the preliminary operations needed to carry out an experimental test. A detailed description of the PIV processing parameters and the thermal calibration procedure will be given. Then, the last part of this section will focus on the time settings and the relative positioning of PIV and Infrared frames with the application of an *ad hoc* calibration plate.

3.1 Experimental arrangement

A schematic of the experimental apparatus is reported in figure 3.1. The facility includes two tanks filled with water, used as working fluid. The small tank, of capacity equal to 25L, is made of stainless steel and it is filled with hot water (298K), while the big one, of capacity equal to 160L, is made of transparent polymethyl methacrylate (PMMA) filled with cold water (293K). The temperatures of the two tanks are measured with two RTD Sensors Pt100 (Omega P-M-1/10-1/8-6-0-p-6) with accuracy equal to 1/10 DIN. A swirl generator device (S) is suspended at a fixed distance H equal to $2D$ from the impingement plate which is a circular sapphire window (diameter 150mm, thickness 4mm) sealed on the bottom of the cold tank. On the dry side of the window, four high-speed PIV cameras (C_i, $i = 1, \ldots, 4$) and a high-speed IR camera (T_1) are vertically positioned.

The test starts when the geared pump (P) pushes the hot water through a flow meter (F). The pump UP3-R is driven by a 24V powered MD10C motor driver controlled by a PWM signal. The hot fluid finally exhausts in the test tank through the swirling nozzle at a constant flow rate equal to $0.15\text{m}^3/\text{h}$, measured with an accuracy of $10^{-3}\text{m}^3/\text{h}$, and Reynolds number equal to 5100.

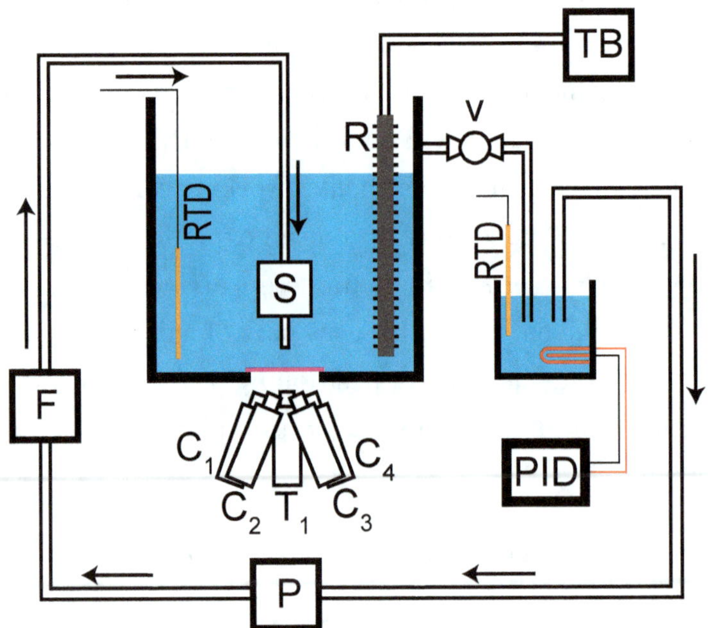

Figure 3.1: Scheme of the experimental apparatus. F) Flow meter; P) Gear type pump; R) Radiator; S) Swirl generator; TB) Thermal Bath; V) Valve; Two RTD Sensors Pt100 are in orange colour. The optical access is in magenta colour; C_i) High-speed PIV cameras; T_1) High-speed IR camera.

For the present experiments water is chosen as working fluid. The choice of $\Delta T = 5\text{K}$ between the two tanks is justified by the non-linear behaviour of water density depending on its temperature. Under these conditions, the hot water issued in the test tank will differ in density of

0.1% from the the cold water. Hence, buoyancy forces can be neglected.

The temperature in the larger tank is set with a Thermo Neslab RTE-210 Refrigerated Bath and a radiator vertically positioned. The smaller tank temperature is set with a self-tuning closed-loop PID controller which controls a submerged heater.

The experimental test procedure requires preliminary operations before the acquisition phase. First of all, the valve is closed and the two target temperatures are reached in the two tanks. Then, the pump is set to work for 90 seconds without triggering the acquisition system. After such a delay, the simultaneous acquisition is triggered with the pump still working. This procedure assures to obtain a quasi-steady phenomenon. Furthermore, in order to avoid any unsteadiness caused by thermalisation of the path from the hot tank to the nozzle, the entire path is coated with neoprene rubber.

The core of the experimental apparatus is the swirl generator device. This consists of three components (figure 3.2):

(a) (inlet) a reinforced flat plate equipped with evenly spaced eight holes and a hose connector attached to;

(b) a disk provided with evenly spaced vanes: this is the component which actually generates the swirling motion;

(c) (outlet) the housing of the swirl generator which acts also as a nozzle.

The whole device is 3D-printed with standard resin which appears as a translucent material with a yellowish colour. It is obtained by liquid resin hardened by a laser which makes it insoluble.

The component committed to welcome the flow is characterised by four spacers which keep in position the swirl generator. The cross-wise section enlargement causes a rapid slow down of the flow, the impact promotes turbulence and facilitates the flow to forget all asymmetries or upstream disturbances.

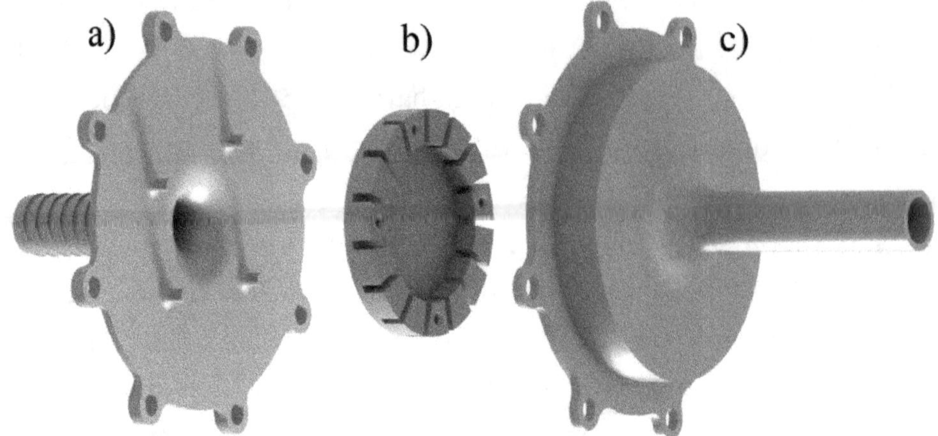

Figure 3.2: Main components of the swirl generator: a) inlet; b) swirl generator; c) nozzle outlet.

The swirl generator is specifically designed in order not to vary the cross-wise section given by the inlet. In other words, the inlet and the outlet areas are equal to the sum of all the side passing-through sections provided by the swirl generator (figure 3.2b). The four holes on the swirl generator's vanes' side (figure 3.3b) act as positioning references on the housing in order to locate the swirl generator always in the same position with respect to the outlet. This last component acts like a nozzle whose diameter is equal to 10mm. Between the inlet and the outlet components, an O-Ring is employed to avoid leakage.

The design parameter suitable to be varied to obtain different swirling ratio is the ϑ angle (figure 3.3b). Hence, nine swirl generators are designed varying the vane angles from 0 to 20deg with a step of 2.5deg.

In the present experimental apparatus, the illumination volume is provided by a high-frequency solid-state diode-pumped LDY303 Nd:YLF laser (2×20mJ). The laser beam is expanded in order to illuminate a volume which is $0.8D$ high along the nozzle-to-wall direction and up to $8D$ along the perpendicular to the nozzle-to-wall direction. The light scattered by the neutrally buoyant LaVision Polyamide High Quality particles (60μm of diameter) is recorded by a tomographic system composed of four

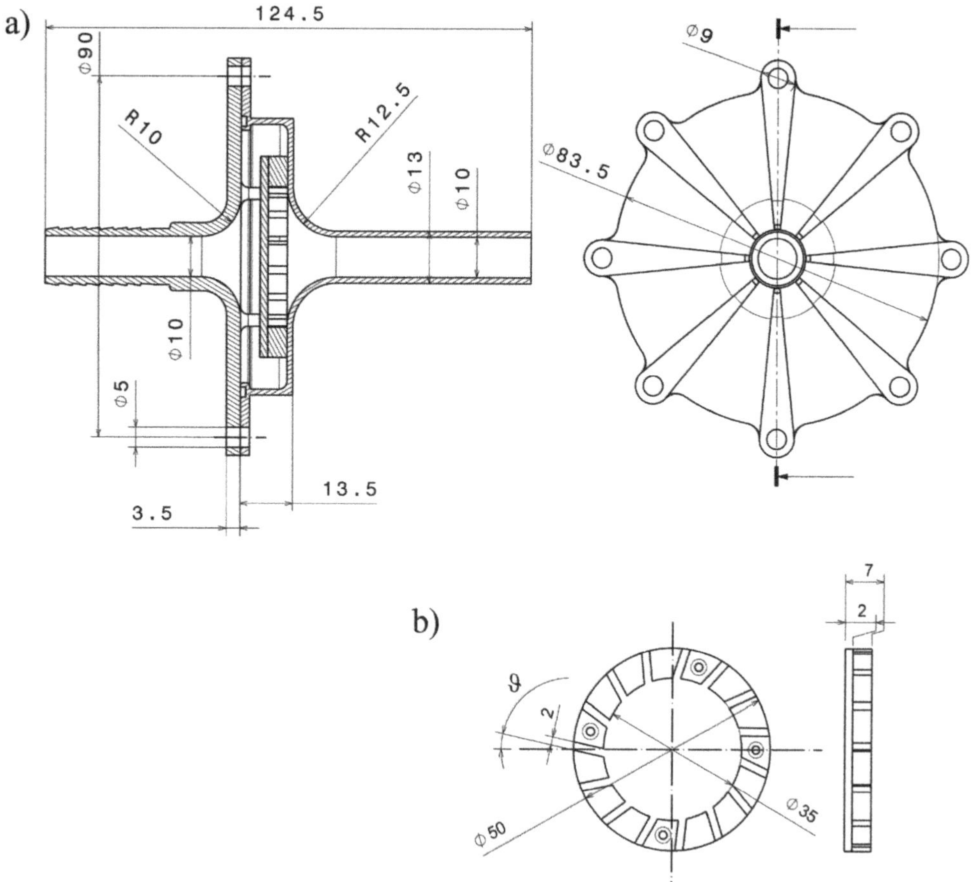

Figure 3.3: Technical drawings of the swirl device: a) assembly of the swirl generator; b) details of the swirl generator.

Speed Sense M110 CMOS cameras (sensor dimensions 1280×800 pixels, pixel size 20µm) arranged vertically below the hot tank in a cross-like configuration. Optics are mounted on cameras through by in-house designed 3D-printed Scheimpflug. Tokina AT-X M100 PRO D objectives of 100mm focal length are used and set with a numerical aperture $f_{\#} = 32$. The resulting field of view is $5.1D \times 3D \times 0.7D$ with a digital resolution of 22.5 pixels/mm.

A CEDIP JADE III (Mid Wave) IR camera working in the $3 - 5$µm band is employed to measure the temperature through the window with a spatial resolution of 2.16 pixels/mm. The camera is positioned perpendicularly to the window with an accuracy of 0.1deg. The camera sensor

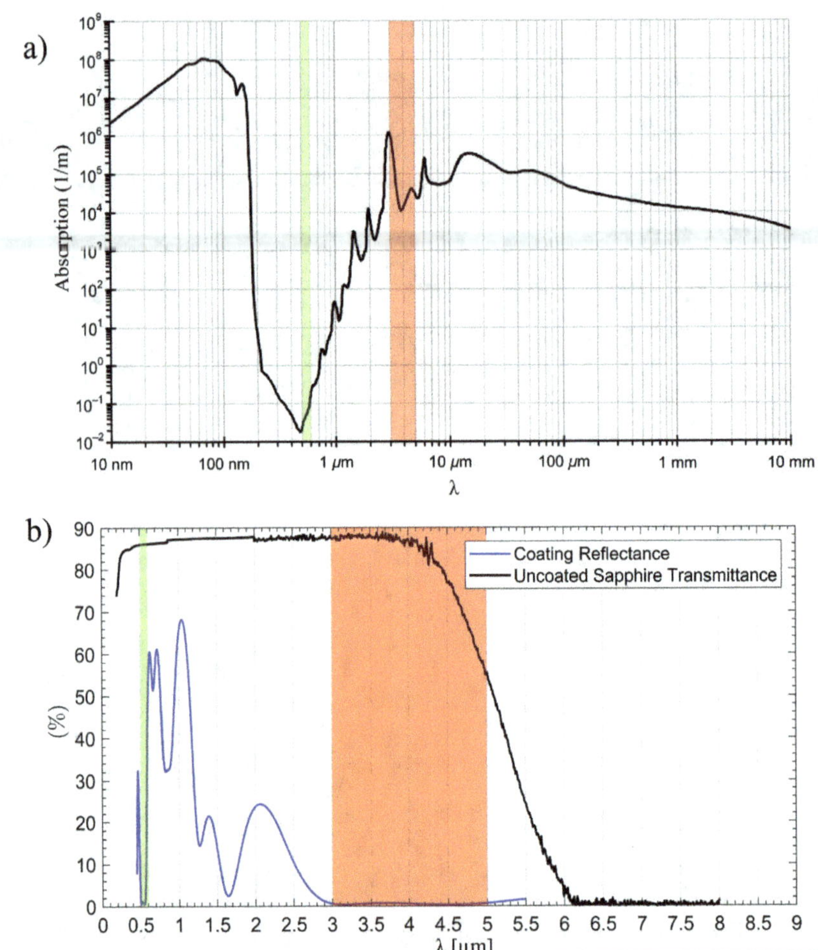

Figure 3.4: a) Absorption spectrum of liquid water (Prahl, 2018); b) Reflectance of anti-reflective coating of the sapphire window (blue solid line); Transmittance of the uncoated sapphire window (black solid line) (Thorlabs, 2018). Nd:YLF laser wavelength range highlighted in green, MWIR range highlighted in red.

has a nominal dimension of 320×240 pixels with a pixel size of 30μm and a temperature sensitivity of 25mK. For the current application the sensor is cropped to 160×120 pixels to achieve a higher sampling frequency of 400Hz. The resulting field of view is $7.3D \times 5.5D$ and the integration time (IT) is set to 1100μs.

The developed setup is based on the property of water to absorb radiation at sub-millimetre scale while it is perfectly transparent in the laser wavelength range (respectively red and green ranges in figure 3.4a).

The dependence of the absorption coefficient of water on the radiation wavelength suggests that the depths of investigation of velocity and of temperature measurement techniques are significantly different. The velocity measurement is performed in a rectangular volume closely located to the impingement wall. On the other hand, the thermal measurement has to be intended as the integration of the electromagnetic radiation emitted from the layer of fluid in contact with the optical window having thickness of the order of the inverse of the absorption coefficient of water in the IR wavelength range (figure 3.5).

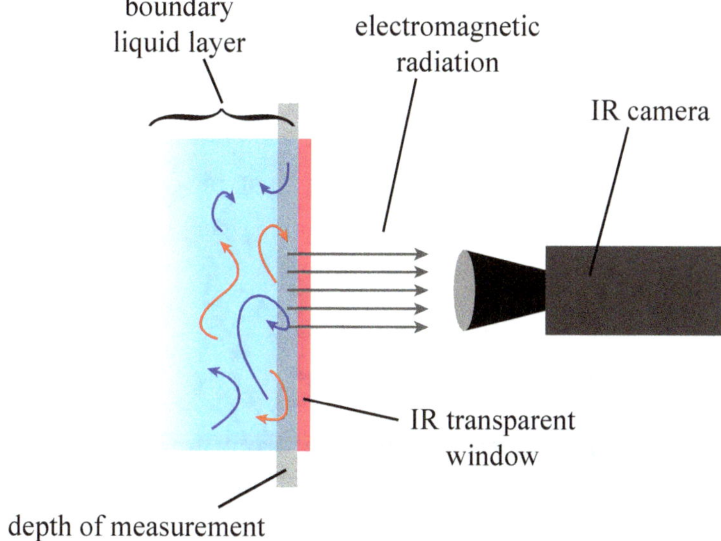

Figure 3.5: Not-in-scale sketch of the depth of investigation of thermal fluctuations in a water facility through a IR transparent window (Znamenskaya et al., 2016a).

The radiation emitted by such a thin layer is measured by the high-speed IR camera. Thus, an IR transparent optical window is needed. The most appropriate IR transparent material for water submerged applications is sapphire. Sapphire optical windows are suitable for water applications because they are not soluble. In addition, such windows are suitable for a large range of wavelengths (UV, visible, and IR wavelengths) with high transmission coefficients (figure 3.4b).

An anti-reflective coating is applied on the dry side of the window. In particular the coating has close to zero reflectance for wavelength included in the Nd:YLF laser range (around 527nm) and in the Mid-Wave Infrared range (3-5µm).

Indeed, exploiting the dependence of water absorption coefficient on the radiation wavelength, the setup described above allows to relate the non-isothermal turbulent fluctuations to the thin thermal boundary layer adjacent to the sapphire window (Bolshukhin et al., 2015; Bolshukhin et al., 2014; Znamenskaya et al., 2016b).

3.2   Data processing

The images acquired with the two applied techniques require a specific processing to be suitable for the image post-processing. In the following, the PIV process parameters and the correction procedure applied to thermal images are described in detail.

Particle Image Velocimetry process

A three-dimensional mapping function from image-space to physical object-space is generated by imaging a calibration target. The initial experimental errors due to system calibration are estimated at approximately 0.5 pixels by the disparity vector field. The misalignment is reduced to less than 0.05 pixels making use of the self-calibration technique (Wieneke, 2008). The raw images, characterised by a uniform $ppp = 0.01$, are pre-processed with subtraction of the minimum intensity at each pixel for the entire sequence, followed by subtraction of the local minimum over a kernel of 7×7 pixels and application of a Gaussian filter with $\sigma_G = 0.5$.

The time-resolved motion analysis employs the most recent algorithms for particle motion tracking and consists of two steps. First, the volumetric light intensity reconstruction of a small set of snapshots is performed. The reconstruction procedure includes the application in sequence of multiple iterations of diffusive simultaneous multiplicative algebraic reconstruction technique (SMART) (Atkinson and Soria, 2009) and diffusive camera-simultaneous multiplicative algebraic reconstruction technique (C-SMART) algorithms through a time-marching SMTE (Lynch and Scarano, 2015) algorithm. This provides a velocity predictor for the next step. The three-dimensional particle field motion is computed by volume deformation iterative multi-grid technique (Discetti and Astarita, 2010; Discetti and Astarita, 2012) with an interrogation box size decreasing from $96 \times 96 \times 96$ to a final box size of $64 \times 64 \times 64$ voxels, including on average, 12 particles. The size of the final interrogation boxes enables a spatial resolution of $0.297D \times 0.297D \times 0.297D$. In order to increase of the spatial sampling, an overlap factor of 75% between adjacent interrogation boxes is chosen, leading to a vector pitch of $0.074D$. The volume under investigation of approximately $5.1D \times 3D \times 0.7D$ is discretized with $535 \times 960 \times 190$ voxels applying a pixel to voxel ratio of 1. The use of 4-camera tomographic system enables 3D object reconstructions with a reconstruction quality Q_f above 0.85 (Elsinga et al., 2006). The second step of the image processing procedure includes the application of the 'Shake-the-Box' method (Schanz et al., 2016) for particle tracking. Then, the particle triangulation is performed by the image matching procedure provided by the IPR technique (Wieneke, 2012). As it can be noticed, the methodology presented is an hybrid between the classic T-PIV tools and the time-resolved Particle Tracking Velocimetry (4D-PTV) approaches. Because of the significant saving of time provided by the latter technique, most of the images are processed with this last approach.

The output of the above process consists of the particle tracks. The interpolation of particle velocities is performed by a least-squares polynomial fitting of local data on a structured Cartesian grid. Hence, for each point of the grid, the particles included in a fixed search radius are identified and the velocities used to define a local polynomial fitting function which is then evaluated at the location of the grid point. In order to reduce the detrimental effects of the ghost particles on the estimation of the velocity field, only particles with trajectories longer than a fixed number of instants are used in the above procedure. For the present experiments, the search radius is set to 48 voxels (about 2mm), a second-order polynomial fitting function is employed and only particles with trajectories longer than 5 time instants are used to estimate the velocity.

Thermal data correction

In order to correctly remove the camera reflection from the Infrared snapshots, called *Narcisus effect*, a thermal calibration procedure is needed. Background images of the field of view are acquired setting the thermal bath at two water temperatures (293K and 298K) and for five camera temperatures (from 309K to 313K, proceeding with 1K steps). By the application of a linear regression procedure over the five points associated to each pixel of the thermal image, two interpolated background images at each test camera temperature are estimated. A linear correction is operated thus providing the instantaneous absolute temperature map:

$$T_{\mathrm{cal}} = \frac{T_{\mathrm{not\ cal}} - T_C}{T_H - T_C} \cdot \Delta T + T_C \tag{3.1}$$

In equation (3.1) , T_{cal} is the resulting temperature map from the linear background correction, $T_{\mathrm{not\ cal}}$ is the temperature map before the application of the linear background correction, T_C and T_H are the background images respectively at lower and higher temperature in the selected range, ΔT is temperature difference between the boundaries of the selected range of temperature (i.e. $T_H - T_C$).

3.3   Time settings and PIV-IR frames spatial correspondence

In order to estimate the unsteady spatial correlation between the two measured quantities, their simultaneous acquisition is needed. Moreover, it is of fundamental importance to know the relative position of the PIV volume over the thermal frame. Hence, in the following the synchronisation settings and the design of a custom target visible in visible and IR range will be described.

3.3.1 Synchronisation settings

The thermal camera, the high speed PIV cameras and the laser share a common trigger provided by a BNC-575 signal generator.

The exact timings and delays are reported in figure 3.6:

- $t_1 = a_1 + a_2 + a_3 = 386.9\mu s$ where

  - $a_1 = 6.5\mu s$ corresponds to the delay between the trigger signal and the laser light emission;

  - $a_2 = 75\mu s$ corresponds to delay needed to center the integration time of the thermal camera (IT $= 1100\mu s$) between the two laser pulses separated by $dt = 1250\mu s$;

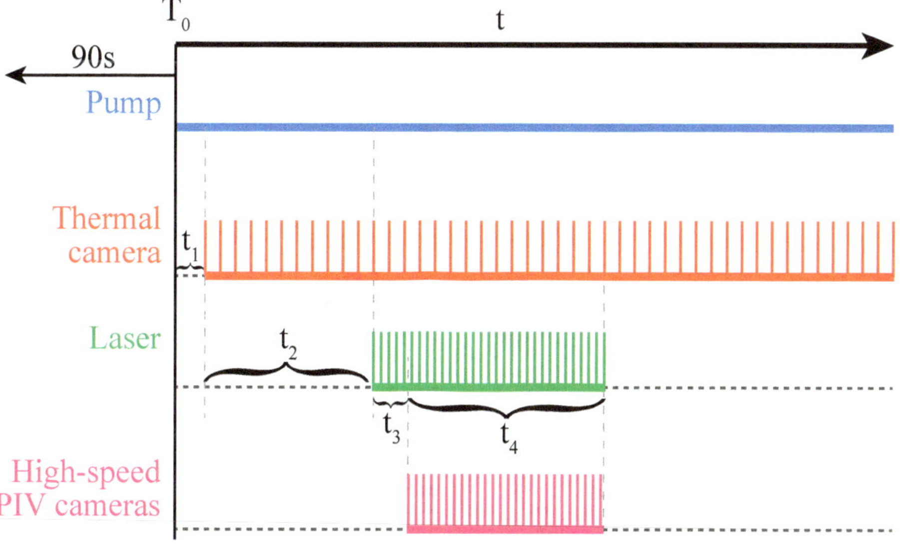

Figure 3.6: Schematic of the time settings adopted for the PIV and the IR acquisition sub-systems.

    ◦ $a_3 = 305.4\mu s$ is the delay between the camera shutter opening and the laser pulse;

- $t_2 = 37.5s$ corresponds to the warm-up of the thermal camera;

- $t_3 = 1s$ corresponds to the laser warm up;

- $t_4 = 5s$ is the duration of the simultaneous acquisition.

The simultaneous acquisition through by the two acquisition systems is carried out at two different frequencies. The PIV subsystem acquires 4000 images at $f_P = 800Hz$ in single frame exposure. The thermal subsystem acquires 48 000 images at $f_T = 400Hz$.

3.3.2　Relative positioning

Since the resolutions of the velocity and the thermographic frames are much different one from each other, this involves two different fields of view (figure 3.7) which need to be correctly overlapped with high accuracy. The relative positioning is obtained through the use of an *ad-hoc* calibration plate (figure 3.8). The calibration target is a PMMA disk

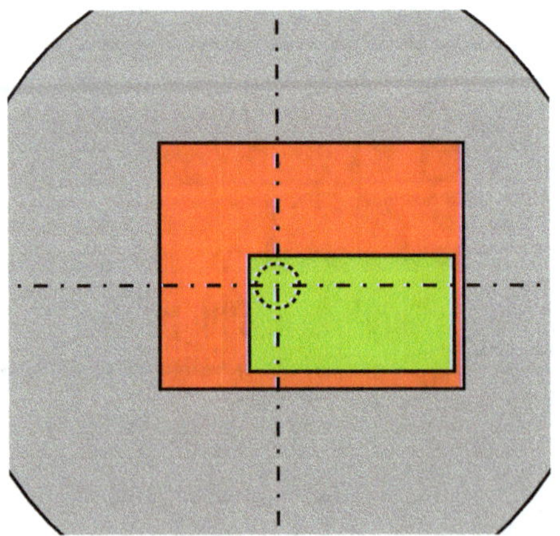

Figure 3.7: Projection of the PIV frame (in green) and of the thermal frame (in red) over the circular sapphire window (in grey). The dashed circle is the projection of the nozzle diameter on the sapphire window.

of 5mm of thickness and 70mm in diameter. Five holes with variable diameter along the depth are located in specific locations.

The drilled disk is sealed from both sides with an adhesive film. From the narrower holes side the film is printed with the same 5-dots pattern, from the other side is not. These five dots are visible from PIV cameras, while, in order to obtain these points also in the IR camera frame, the air inside the holes is heated by a heat gun from the opposite side of the sapphire window. In this way, these dots can be also observed by the IR camera. The target is placed on the sapphire window. Then, once acquired the target pattern with the two acquisition systems, the coordinates of such points now obtained in both frames are determined in the object-space reference system with a minimisation technique.

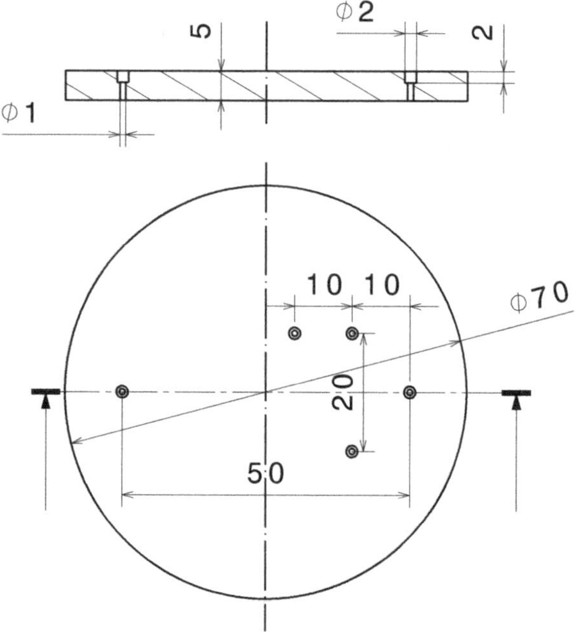

Figure 3.8: Technical drawings of the target plate adopted to obtain the relative positioning of the PIV and the thermal frames.

3.4 Test matrix and reference coordinate systems adopted

In the next chapter two main analyses are performed with the same operating conditions: volumetric flow rate equal to $0.15\text{m}^3/\text{h}$, Reynolds number $Re = 5100$.

First, the flow field of a free isothermal swirling jet is discussed. For each of the swirl generators designed, 4000 images at frequency of 800Hz in single frame mode are acquired.

A laevorotatory coordinate system is employed in such analyses. It has the origin located on the nozzle exit and the x axis, directed downward, is aligned with the nozzle axis. The investigation volume extends among the following boundaries:

$$\left\{x^N \times y^N \times z^N\right\}/D^3 =$$
$$= \left\{\left[x^N_{LB}, x^N_{HB}\right] \times \left[y^N_{LB}, y^N_{HB}\right] \times \left[z^N_{LB}, z^N_{HB}\right]\right\}/D^3$$
$$\in \left\{[0.12, 0.75] \times [-1, 1] \times [-1, 1]\right\}$$

according to the reference system reported in figure 3.9. Such a volume is chosen in order to be able to evaluate the swirl numbers and the mean flow field characteristics of the investigated devices.

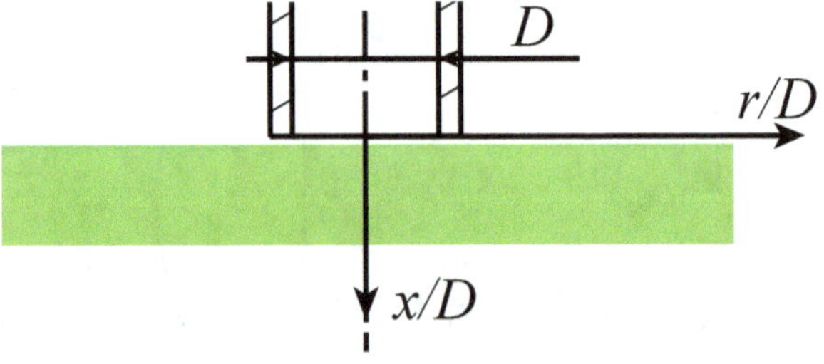

Figure 3.9: Reference coordinate system adopted and relative positioning of the laser volume (in green) for the free flow field investigations.

In the second part of the next chapter, the investigation of the temperature and of the velocity fields of an hot impinging swirling jet suspended at $H/D = 2$ from a flat wall is performed for five swirl numbers. For each of the investigated swirl ratios, the PIV acquisition frequency is set to 800Hz in single frame mode and the thermal acquisition frequency is set to 400Hz acquiring respectively 4000 and 48000 images for each experimental test.

The laevorotatory coordinate system adopted in such analyses has the origin located on the impingement wall and the x axis, aligned with the nozzle axis, is directed from the wall to the nozzle (figure 3.10). The measurement boundaries extracted from the illuminated volume are the following:

$$\left\{ x^W \times y^W \times z^W \right\}/D^3 =$$
$$= \left\{ \left[x^W_{LB}, x^W_{HB} \right] \times \left[y^W_{LB}, y^W_{HB} \right] \times \left[z^W_{LB}, z^W_{HB} \right] \right\}/D^3$$
$$\in \left\{ [0.02, 0.73] \times [-0.71, 4.18] \times [-2.19, 0.44] \right\}.$$

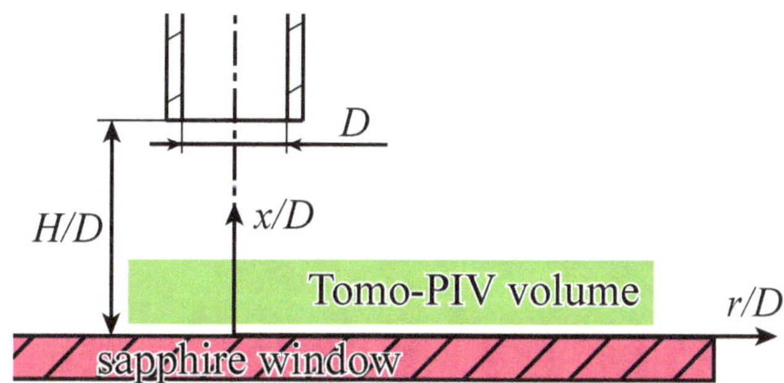

Figure 3.10: Reference coordinate system adopted and relative positioning of the nozzle and of the laser volume (in green) over the sapphire window (in magenta) for the impinging experimental tests.

The investigation volume presented above is chosen in order to be able to evaluate the flow field characteristics in proximity of the wall simultaneously to the wall temperature behaviour caused by the impinging of the hot swirling jet.

3.5   Estimation of the swirl number

The swirl number is a non-dimensional quantity defined as the ratio between the axial flux of the azimuthal momentum G_θ and the axial flux of the axial momentum G_x multiplied by the equivalent nozzle radius $D/2$ (Gupta et al., 1984). As a matter, there exist many definitions of the swirl number depending on the difficulty in measuring the quantities involved and depending on the assumptions available regarding the specific flow field (Toh et al., 2010).

In order to evaluate the swirl number, the investigation volume is that reported in figure 3.9. In the present work, the most general expression to evaluate the swirl number will be applied:

$$S = \frac{2\pi\rho \int_0^{D/2} r^2 \left(UW + \overline{u'w'} \right) dr}{2\pi\rho \left\{ \int_0^{D/2} r \left[U^2 - \frac{1}{2}W^2 + \overline{u'^2} - \frac{1}{2} \left(\overline{w'^2} + \overline{v'^2} \right) \right] dr \right\} \cdot D/2} \tag{3.2}$$

where the capital letter are the time-averaged quantities while the small ones with the apex indicate the velocity fluctuations ($u \triangleq U + u'$). Additional details about the swirl number definition are reported in appendix A.

4 | Results

In the present chapter the main results will be presented. If not differently declared, the coordinate system will be a cylindrical polar coordinate system. Hence, the velocity components in (x, r, θ) will be named (u, v, w) with the first direction aligned with the nozzle direction.

The chapter is organised as follows. The first section focuses on the flow field in proximity of the nozzle. The time-averaged profiles are discussed for each one of the designed swirl generators. The circulation distributions are also presented. Then, five swirl numbers are selected $(S = 0, 0.23, 0.43, 0.61, 0.74)$ and the related three-dimensional instantaneous patterns are described. Among them, $S = 0.61$ is selected to apply Proper Orthogonal Decomposition technique (Berkooz et al., 1993) to identify the main features of a swirling jet (Lumley, 1967).

The second section focuses on the flow field of a swirling hot jet impinging on a flat wall. In the first part, the time-averaged velocity and temperature profiles are discussed. The second part focuses on the instantaneous thermal and velocity fields. Then, the time sequences of temperature fluctuations are presented for three distances from the impingement centre. The maps are discussed and the mixing performances are compared. In order to estimate the time-averaged radial velocity profiles at the wall, the thermal fluctuation snapshots are processed with a cross-correlation algorithm. The instantaneous characterisation of the two measured quantities is completed with the description of three-dimensional velocity snapshots overlapped to the corresponding thermal fluctuation maps. In the third part POD technique is applied to both quantities.

Last section is focused on the discussion of the relationship between velocity and temperature fluctuation distributions. The first analysis presented is the Extended POD modal decomposition technique. Then, the time-averaged correlation between the last slice of velocity domain and temperature footprint is investigated.

4.1 Flow field of a free swirling jet

The experimental measurements in proximity of the nozzle are acquired with the swirl generator positioned far from the wall and the laser volume located in proximity of the nozzle according to the schematic reported in figure 3.9. No temperature difference is set between the jet and the ambient.

In the following, in order to distinguish the different swirl generators, all results will be referred to the swirl number instead of the vane angle. The correspondence between the designed vane angle and the swirl number generated is reported in table 4.1. The evaluation of S is strictly dependent to the radial velocity profiles presented in the following.

Vane angle [deg]	Swirl number [-]
2.5	0.08
5.5	0.18
7.5	0.23
10.0	0.31
12.5	0.43
15.0	0.51
17.5	0.61
20.0	0.74

Table 4.1: Correspondence between the designed vane angle and the swirl number estimated with equation (3.2).

4.1.1 Radial distributions and swirl number estimation

In figure 4.1 the radial velocity distributions of all swirl generators at distance $x/D = 0.12$ from the nozzle exit are reported. The bulk velocity U_0, obtained from the flow rate and the nozzle diameter, is assumed as reference velocity.

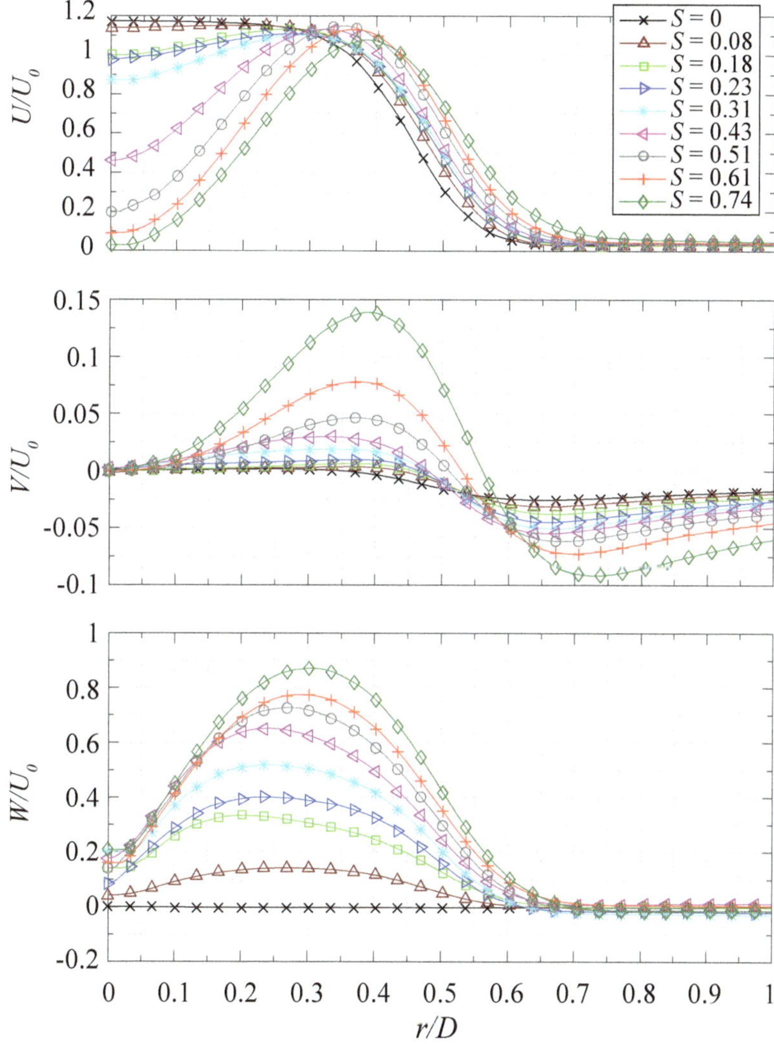

Figure 4.1: Radial velocity distributions of a free submerged swirling jet ($x/D = 0.12$). Legend reports the swirl numbers.

The radial profiles of the axial velocity component provide several significant information. The no swirling case has a smooth top hat shape. The maximum of velocity is reached in the range $0 < r/D < 0.3$. For radial distance larger than $r/D = 0.3$ the axial velocity component rapidly drops reaching the zero value at $r/D = 0.7$. It is worth noting that the no swirling jet does not expand much once emerged from the nozzle. The distribution of V is approximately flat and close zero from the origin to $r/D = 0.4$. Then, the radial velocity component experiences a slight decrease in value from $r/D = 0.4$ to $r/D = 0.65$ reaching the minimum value of $V/U_0 = -0.025$. After such a minimum, the distribution of V shows a weak increasing trend.

The introduction of a non-zero vane angle in the swirl generator provides a swirl number equal to 0.08. The azimuthal component of velocity vector is noticeable and the maximum value, equal to $0.15U_0$, is located at $r/D = 0.25$. A less predictable consequence of the rotation is the decreasing of the axial velocity along the jet axis from $1.2U_0$ to $1.15U_0$. This affects the whole velocity profile to satisfy the mass conservation. As a matter, the velocity distribution slightly widens.

With the increase of the swirl number, each component of the velocity vector is affected in a different manner:

- the axial velocity profiles experience a decrease along the jet axis $(r/D = 0)$ and, as a consequence, the distributions widen. For the largest swirl number the axial velocity is close to experience a reverse flow;

- the radial component of velocity rapidly increases after that the swirl number has overcome the value $S = 0.43$. This results in a fast widening of the cross-wise jet section which affects also the axial component of velocity vector distribution;

- simultaneously with the increasing of the azimuthal velocity component the maximum of velocity moves from $r/D = 0.15$ to $r/D = 0.3$.

In figure 4.2 the radial distributions of the Reynolds stresses are reported. The distributions related to the no-swirl case are flat and close to zero value except for $\overline{u'^2}$ which experiences a significant increase at $r/D = 0.45$. This can be ascribed to the shear layer.

A slight increase of the swirl number $(0.08 < S < 0.18)$ affects $\overline{u'^2}$, $\overline{v'^2}$ and $\overline{w'^2}$ in proximity of the jet axis. The second-order moments $\overline{u'v'}$ and $\overline{v'w'}$ are not influenced by the introduction of tangential motion, while the last one, $\overline{u'w'}$, is dominated by the u' contribution presenting a slight decrease in correspondence of the nozzle.

Increasing the swirl number, all the Reynolds stresses are affected and experience a profile broadening and the increasing of their absolute values.

For swirl numbers greater than $S = 0.43$, the $\overline{u'^2}$ experiences a second local maximum located at $r/D = 0.15$. With the increasing of the swirl number, the maximum location moves away from the origin and oscillates between $r/D = 0.1$ and $r/D = 0.2$. The same behaviour can be recognised in $\overline{v'^2}$ and $\overline{w'^2}$ radial distributions but with less evidence with respect to the $\overline{u'^2}$ case.

The mixed second-order moments have a more complex behaviour. In particular, the $\overline{u'v'}$ distribution is characterised by a minimum located at $r/D = 0.2$ which increases in absolute value with the increasing of the swirl number. A local maximum located at $r/D = 0.5$ is also present but it is less evident.

The $\overline{v'w'}$ distribution is characterised by the presence of a minimum and a maximum. They are respectively located at $r/D = 0.3$ and $r/D = 0.5$. The former can be ascribed to the drag of the external quiescent flow field. The latter is due to the forced swirling motion imposed by the jet.

The last mixed second-order distribution $\overline{u'w'}$ takes some from $\overline{u'v'}$ and $\overline{v'w'}$ behaviours but with some differences. A maximum located at $r/D = 0.5$ is present. A significant difference with the other mixed second-order moments is the strongly dependent behaviour on swirl number, in particular for swirl ratios greater than $S > 0.23$. As a matter, a second maximum appears at $r/D = 0.1$ while the local minimum de-

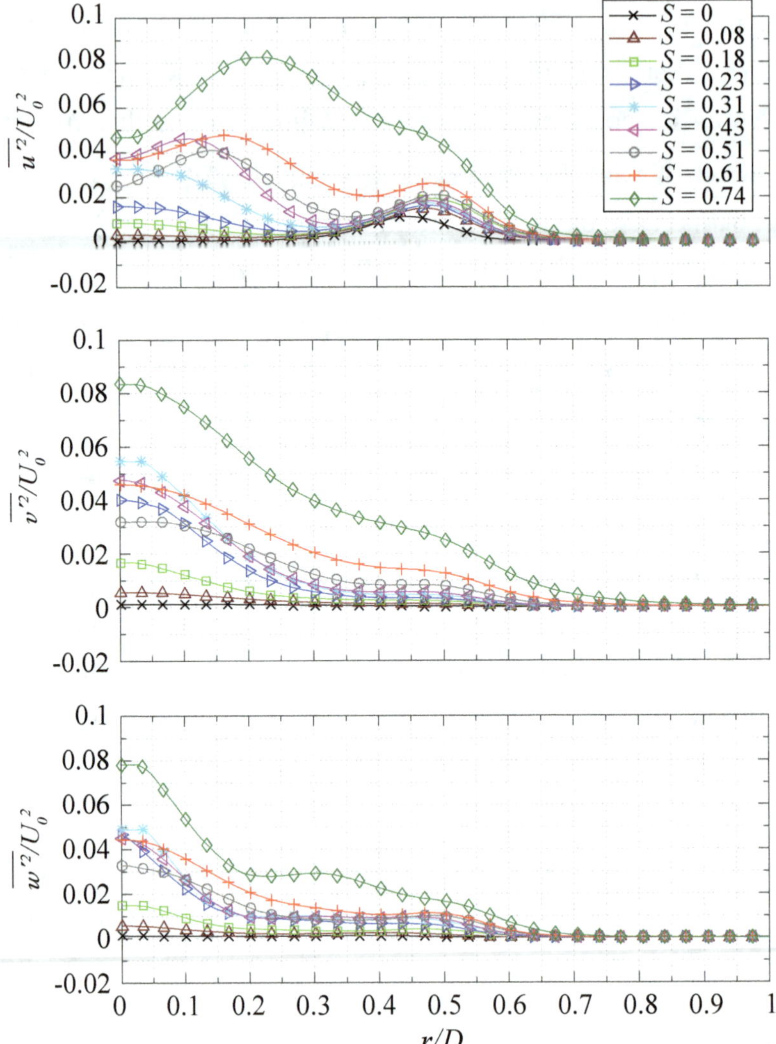

Figure 4.2: Radial distributions of Reynolds stresses for a free submerged swirling jet $(x/D = 0.12)$. Legend reports the swirl numbers.

creases in value and moves from $r/D = 0.2$ to $r/D = 0.45$.

Particularly interesting is the radial distribution of circulation Γ estimated from the time-averaged velocity components (figure 4.3). Such a quantity is calculated according to the following expression:

$$\Gamma(r) = \oint_0^{2\pi} W r\, d\theta \qquad (4.1)$$

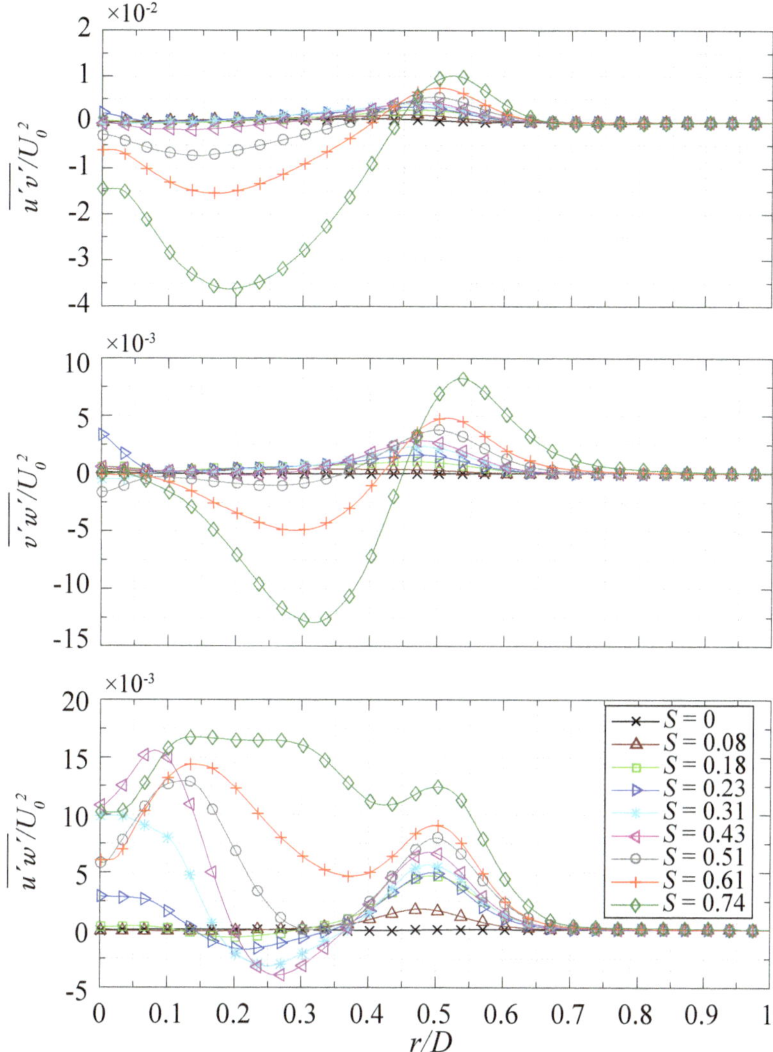

Figure 4.2 (Cont.): Radial distributions of Reynolds stresses for a free submerged swirling jet ($x/D = 0.12$). Legend reports the swirl numbers.

where W is the time-averaged tangential velocity component, r is the radial coordinate, θ is the azimuthal coordinate used as integration variable and the integral is evaluated for several closed circles of radius included in the range $0 < r < D$. Two trends appear noticeable with the increasing of the swirl number: the increasing of the circulation maximum value and the displacement of the maximum location. The first result is not surprising: the circulation increases as much as the tangential component of the flow field increases. The second result suggests that the swirl number affects also the location of the maximum of circulation. As a matter, the maximum value of circulation is found to range between $0.35D$ and $0.4D$ with the increasing of the swirl number in accordance with H. Liang and Maxworthy (2005).

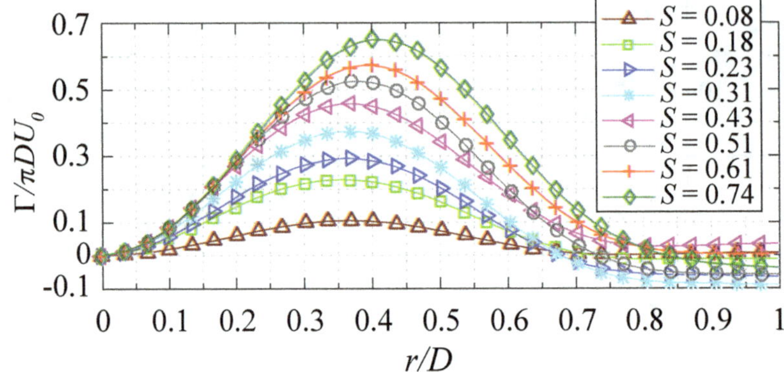

Figure 4.3: Radial distributions of normalised time-averaged circulation at $x = 0.12D$.

4.1.2   Three-dimensional instantaneous flow field

The results presented above suggest a clear trend. Hence, five swirl numbers are selected ($S = 0, 0.23, 0.43, 0.61, 0.74$) to describe the three-dimensional instantaneous free flow field of a swirling jet in proximity of the nozzle. In the following, each row of the reported figures provides two different visualisations of the same snapshot.

The first case we will focus on is the no-swirl case (figure 4.4). The internal nozzle diameter is marked in red. In figure 4.4a two isosurfaces are reported: the grey one is related to $u/U_0 = 1$, the contoured one is obtained by the application of Q-criterion (Hunt et al., 1988) with threshold set to $Q = 0.15$. The contour colour refers to the normalised radial velocity component.

The grey isosurface gradually widens. This suggests that the spreading of the jet is noticeable since the first steps out of the nozzle. The annular-

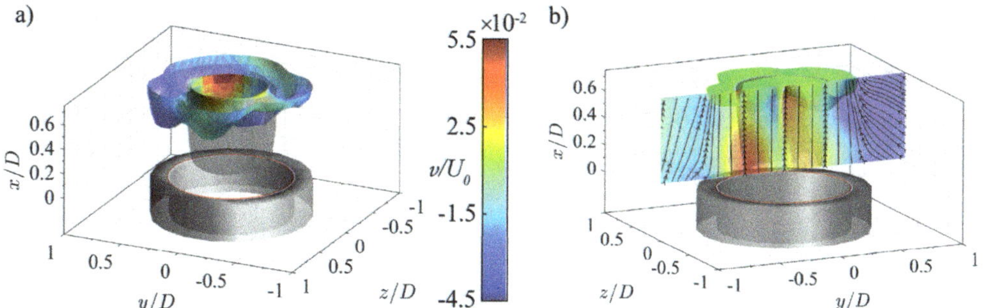

Figure 4.4: Three-dimensional visualisation of the instantaneous flow field in proximity of the nozzle for $S = 0$. a) In translucent grey colour it is represented the isosurface of axial velocity $u/U_0 = 1$. The contoured isosurface corresponds to $Q = 0.15$. The colorbar is related to v/U_0. b) Green and grey isosurfaces correspond respectively to $Q = 0.15$ and $u/U_0 = 1$. The slice located in correspondence of the $z/D = 0$ plane is filled with in-plane black streamlines and the background is contoured with v/U_0.

shaped vortex is a typical pattern recognisable in the flow field emerging from a free round jet. It is located between the core flow and the resting ambient flow thus spotting the shear layer instabilities which commonly generates during the emerging of the jet from the nozzle.

The contoured slice reported in figure 4.4b gives additional information about the snapshot presented in figure 4.4a. The flow field is dissected along the $z/D = 0$ plane and the resulting slice is contoured with the radial component of velocity and the corresponding in-plane streamlines. As it can be noticed, the radial velocity component distribution reported is not axisymmetric, neither at stations close to the nozzle nor at the opposite boundary of the investigation volume. Despite of that, the streamlines are not noticeably affected, suggesting that such an unsteady anisotropy is not strong enough to significantly deflect the main flow. On the other hand, the streamlines and the contour related to regions located out of the nozzle diameter provide an evidence of the entrainment phenomenon which drags the resting ambient fluid towards to the jet axis.

The superimposition of tangential motion to a conventional free round jet introduces several different patterns and dynamics. In figure 4.5 the instantaneous flow field of a swirling jet ($S = 0.23$) is reported. Figure 4.5a reports the Q isosurfaces which are contoured with the radial component of velocity. The farthest plane from the nozzle ($x = x_{HB}^N$) is marked in translucent grey colour. The vector field reported refers to the velocity components parallel to that plane ($\underline{V}_\parallel = \underline{v} + \underline{w}$). The contour colour refers to vectors magnitude. For the sake of colour contrast, the colorbar related to the vector field and to the isosurfaces' contour included in the same sub-figure reports different ranges for the two quantities reported.

In figure 4.5b the same Q-isosurface of figure 4.5a but in green colour and a grey isosurface representative of $u/U_0 = 1$ are reported. Additional information is reported on the slice which dissects the flow field along the plane $z/D = 0$. The slice is filled with the in-plane black stream-lines and the background is contoured with the azimuthal component of velocity. In addition, ten magenta streamlines equally distributed over a circumference of radius equal to $0.1D$ depart from $x = x_{LB}^N$.

Figure 4.5a shows a spiral vortical structure and a cylindrical vortex core. The distribution of radial velocity component is not uniform in the reported snapshot. As a matter, the radial velocity is positive in proximity of the vortex core. This occurrence results in the displacement and in the bending of the vortex core from the jet axis. The vector field reported shows a high in-plane velocity spot displaced from the jet axis.

Figure 4.5b confirms such a conclusion. As a consequence, the stream-lines are clearly disturbed since $x = x_{LB}^N$. That disturbance affects the whole vortex core which is not perfectly aligned to the jet axis showing a slight distortion. The contour on the slice shows that the azimuthal velocity component is not axisymmetric.

The passage from a low ($S = 0.23$) to a moderate ($S = 0.43$) swirl number emphasises some of the patterns previously described. By now, the grey isosurfaces on the left sided images are related to $v/U_0 = 0.1$. The grey isosurfaces on the right sided images are, where present, repre-sentative of a negative axial velocity component ($u/U_0 = -0.1$).

In figure 4.6a a thick spiral vortical structure which turns around a

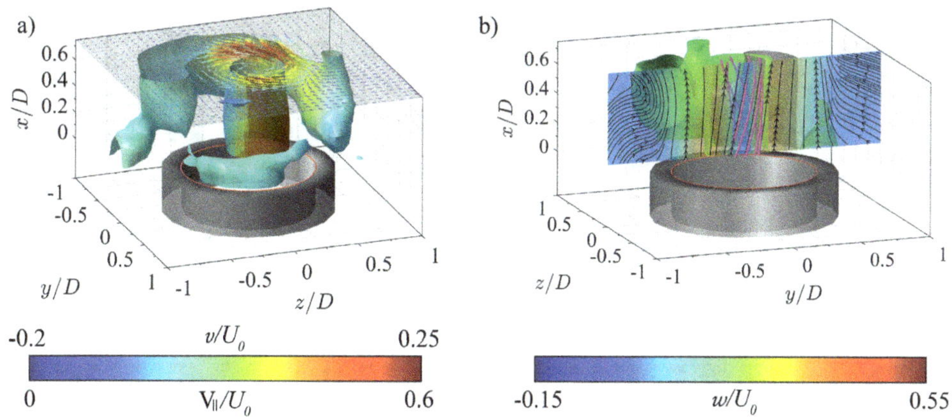

Figure 4.5: Three-dimensional visualisation of the instantaneous flow field in proximity of the nozzle for $S = 0.23$. a) The isosurface corresponds to $Q = 0.1$ and it is contoured with v/U_0. The vector field located on the plane $x = x_{HB}^N$ is contoured with $V_{\parallel}/U_0$. The colorbar is shared between v/U_0 and $V_{\parallel}/U_0$ with different ranges. b) Green and grey isosurfaces correspond respectively to $Q = 0.1$ and $u/U_0 = 1$. The slice located in correspondence of the $z/D = 0$ plane is filled with in-plane black streamlines and the background is contoured with w/U_0. Ten magenta streamlines, equally distributed over a circumference of radius equal to $0.1D$, depart from $x = x_{LB}^N$.

central vortex is reported. Such a structure is coupled with a region of positive radial velocity component. Not surprisingly, the pitch of the helix is greater than the previous discussed swirl number. As it can be noticed from the mostly blue contour colour, the major difference with respect to the previous case is the entrainment enhancement. It is noticeable the effect of the precessing vortex on the vortex core which is starting to deflect with respect to the jet axis.

Additional information can be deduced from figure 4.6b. The $z = 0$ plane dissects simultaneously the vortex core and the outer spiral vortex. As shown in figure 4.6a, the vortex core is slightly bent toward the positive y direction. In addition, the increasing of cross sectional area of the coloured streamlines envelope suggests that there is a smooth spreading of the jet. The asymmetry caused by the vortex core causes an imbalance of the azimuthal component of velocity which is higher in absolute value

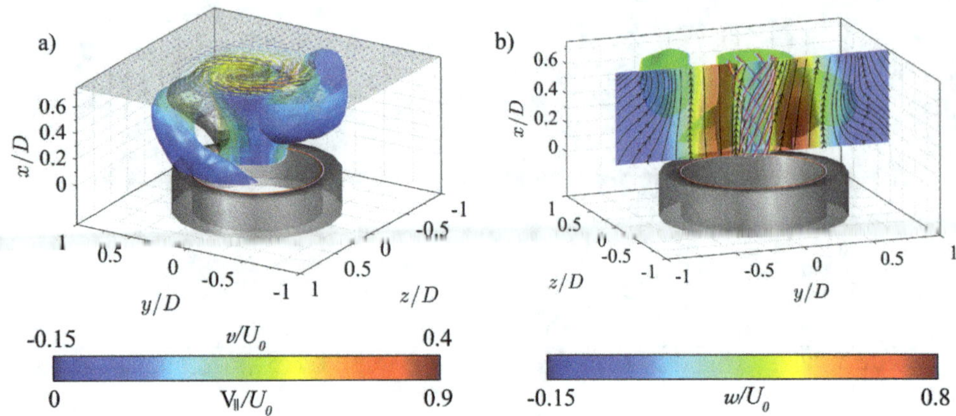

Figure 4.6: Three-dimensional visualisation of the instantaneous flow field in proximity of the nozzle for $S = 0.43$. a) The coloured isosurface corresponds to $Q = 0.1$ and it is contoured with v/U_0. The vector field located on the plane $x = x_{HB}^N$ is contoured with $V_\parallel/U_0$. The colorbar is shared between v/U_0 and $V_\parallel/U_0$ with different ranges. Grey isosurface corresponds to $v/U_0 = 0.1$. b) Green isosurface corresponds to $Q = 0.1$. The slice located in correspondence of the $z/D = 0$ plane is filled with in-plane black streamlines and the background is contoured with w/U_0. Ten magenta streamlines, equally distributed over a circumference of radius equal to $0.1D$, depart from $x = x_{LB}^N$.

on positive y direction.

Figure 4.7 reports four significant instantaneous snapshots related to $S = 0.61$. They show how a moderate swirl number can completely change the organisation of the flow field in proximity of the nozzle. Figure 4.7a and figure 4.7b report the occurrence of an evident spiral vortex located around a thick vortex core. The radial component of velocity is not uniformly distributed. As a matter, a positive radial velocity isosurface is coupled with the spiral. In addition, a negative spot of v is located on the upper part of the central vortex. In this frame the red spot on the vector field at $x = x_{HB}^N$ shows a clear displacement from the jet axis. This coincides with the maximum of in-plane vorticity which experiences a precessing motion around the vortex core at a distance of $0.08D$ from the nozzle axis.

Figure 4.7b gives additional information about the self-organisation of the flow field. The vortex core is noticeably inclined towards the y positive direction and experiences a slight spreading. Despite of that, the axis of the truncated cone shaped envelope of coloured streamlines is still included in the $z = 0$ plane. The streamlines have a constant pitch and do not experience abrupt disturbances.

Figure 4.7c report a highly varying radial velocity component distribution over the vortical structures. As it can be noticed there are two critical spots located along the vortex core at the two extremes of domain extension along x direction. The positive value of radial velocity component is concentrated in proximity of the nozzle while the negative one is located on the upper part of the vortex core. Such a pattern suggests the presence of a precessing spiral motion around the jet axis. The envelope of the coloured streamlines in figure 4.7d tends to follow a spiral path bending along the positive z direction.

Figure 4.7e and figure 4.7f are representative of how much the general flow organisation can change in case of incoming recirculation phenomena. The Q isosurface reported in figure 4.7e is characterised by a double spiral structure. The external one is the usual spiral structure present also in the previous frames (figure 4.7a and figure 4.7c). The internal one can be ascribed to the precessing vortex core (PVC) (Cala et al., 2006). Hence, the previously straight vortex core is, in this case, completely distorted. The positive radial velocity component isosurface can be thus associated as coupled with both spiral structures and remarks the boundary between them. The rest of the flow field is characterised by negative v component resulting in a significantly enhanced entrainment effect.

The slice reported figure 4.7f gives additional information on the organisation of the mutual interaction between the two vortical structures. The black streamlines clearly spot the spiral vortex cut at the two boundaries along the x direction of the investigation volume. Approaching to the vortex core the streamlines curve smoothly. Particularly interesting is what happens within the region included in a cylinder centred on the nozzle axis and of radius equal to $0.25D$. The coloured streamlines appears

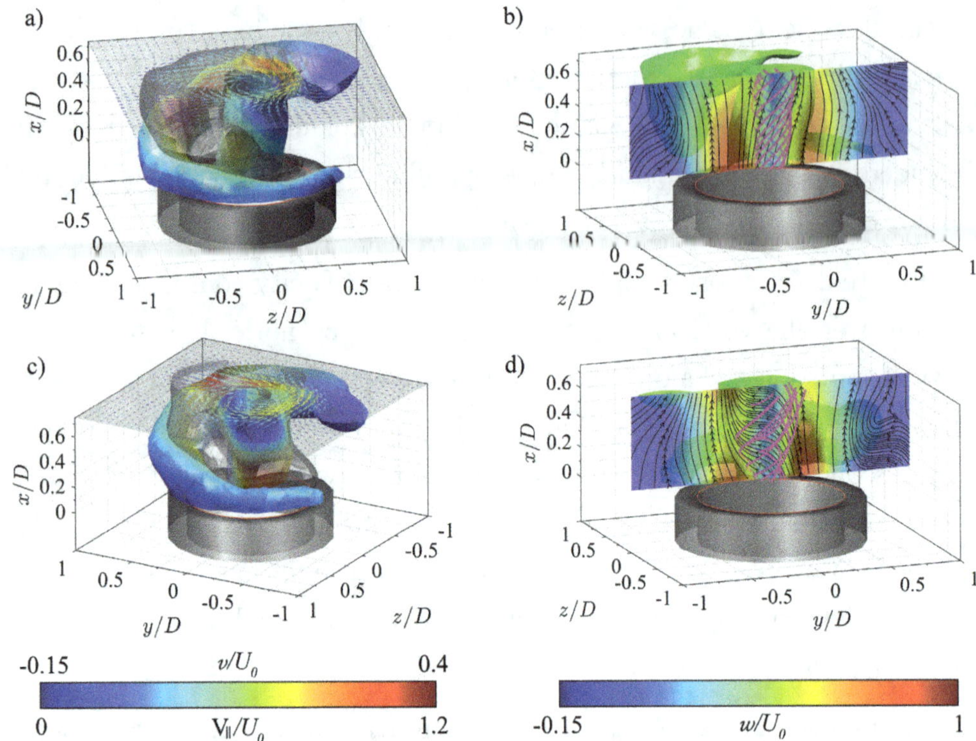

Figure 4.7: Instantaneous measurements of the flow field in proximity of the nozzle for $S = 0.61$. a,c) The coloured isosurface corresponds to $Q = 0.1$ and it is contoured with v/U_0. The vector field located on the plane $x = x_{HB}^N$ is contoured with $V_\parallel/U_0$. The colorbar is shared between v/U_0 and $V_\parallel/U_0$ with different ranges. Grey isosurface corresponds to $v/U_0 = 0.1$. b,d) Green and grey isosurfaces correspond to $Q = 0.1$ and $u/U_0 = -0.1$ respectively. The slice located in correspondence of the $z/D = 0$ plane is filled with in-plane black streamlines and the background is contoured with w/U_0. Ten magenta streamlines, equally distributed over a circumference of radius equal to $0.1D$, depart from $x = x_{LB}^N$.

to follow the internal spiral vortex.

Figure 4.7g and figure 4.7h show the flow organisation in case of a significant recirculation region. The overall organisation of the flow field shown in figure 4.7g does not appear particularly different from other scenarios discussed: an outer spiral vortex embraces the central vortex core.

Figure 4.7h gives information about what it is happening inside of the

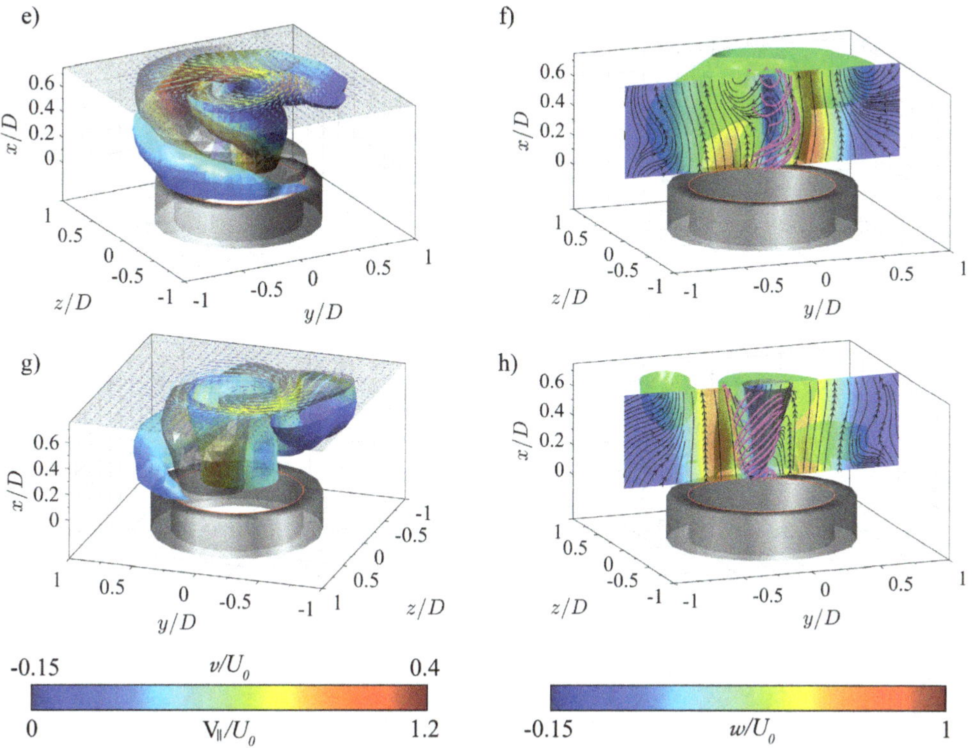

Figure 4.7 (Cont.): Instantaneous measurements of the flow field in proximity of the nozzle for $S = 0.61$. e,g) The coloured isosurface corresponds to $Q = 0.1$ and it is contoured with v/U_0. The vector field located on the plane $x = x_{HB}^N$ is contoured with $V_\parallel/U_0$. The colorbar is shared between v/U_0 and $V_\parallel/U_0$ with different ranges. Grey isosurface corresponds to $v/U_0 = 0.1$. f,h) Green and grey isosurfaces correspond to $Q = 0.1$ and $u/U_0 = -0.1$ respectively. The slice located in correspondence of the $z/D = 0$ plane is filled with in-plane black streamlines and the background is contoured with w/U_0. Ten magenta streamlines, equally distributed over a circumference of radius equal to $0.1D$, depart from $x = x_{LB}^N$.

vortex core. The grey isosurface reported refers to a region with negative axial velocity component equal to $u/U_0 = -0.1$. The recirculation region acts as support for the whole vortical central structure. It has a truncated cone shape which penetrates the whole central core region. The occurrence of such a phenomenon completely changes the usual internal streamline distribution. The black streamlines behave differently depending if they are located outside or inside the nozzle extension. In the outside region they are affected by the presence of the outer spiral vortical structure and their curvature smoothly change approaching to the core. Within the core, the recirculation occurrence completely distorts the streamlines which clearly show two sharp bends. In the region interested by the recirculation, the coloured streamlines give three-dimensional information about the organisation of the flow field.

The last investigated flow field is related to swirl number equal to $S = 0.74$ (figure 4.8). All frames report a recirculation zone each time of different size. In the first row of images (figure 4.8a and figure 4.8b) the outer spiral vortex appears to couple with the vortex core starting to distort it on the upper part. The explanation of such a deformation can be deduced by the dissection of the vortex core reported in figure 4.8b where a small recirculation zone is visible. Hence, the occurrence of a recirculation phenomenon which affects a small region of the domain triggers the precessing vortex core formation. The azimuthal velocity component distribution loses axisymmetry in proximity of the recirculation zone where the w velocity component decreases. In this case, the maximum of the in-plane velocity $V_\parallel$ on the grey plane on the left sided figure is located at $0.1D$ further away from the jet axis.

The second row of images (figure 4.8c and figure 4.8d) reports the occurrence of a larger recirculation region. In this case the external spiral vortical structure is characterised by an extremely varied distribution of the radial velocity component. The highest value of v is reached on the upper part, internal side of the vortex core, thus suggesting the presence of the PVC. A detailed description of such a phenomenon is reported in figure 4.8d. The black streamlines are issued by the spiral vortex. The streamlines located between the vortex core and the outer region show smooth curvature gradually adapting to the internal flow. The coloured streamlines wrap as usual around the jet axis until $x/D = 0.3$. After that station the recirculation region starts to influence the flow field spreading the streamlines path and distorting the regular helical pattern because

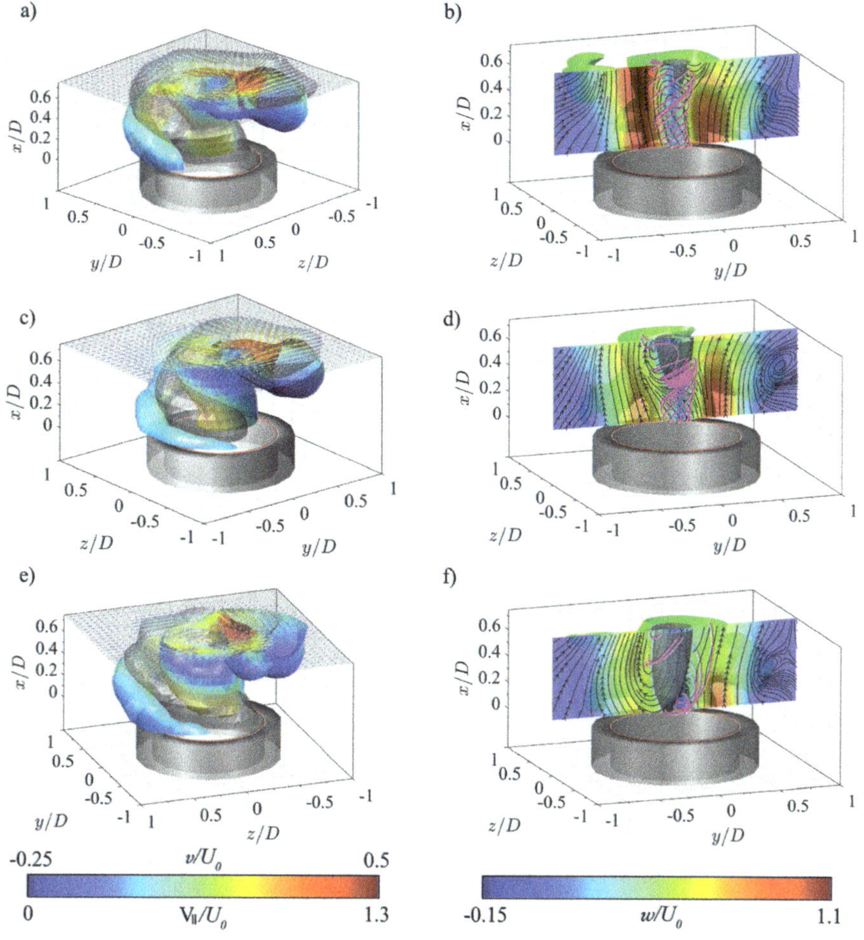

Figure 4.8: Instantaneous measurements of the flow field in proximity of the nozzle for $S = 0.74$. a,c,e) The coloured isosurface corresponds to $Q = 0.1$ and it is contoured with v/U_0. The vector field located on the plane $x = x_{HB}^N$ is contoured with $V_\parallel/U_0$. The colorbar is shared between v/U_0 and $V_\parallel/U_0$ with different ranges. Grey isosurface corresponds to $v/U_0 = 0.1$. b,d,f) Green and grey isosurfaces correspond to $Q = 0.1$ and $u/U_0 = -0.1$ respectively. The slice located in correspondence of the $z/D = 0$ plane is filled with in-plane black streamlines and the background is contoured with w/U_0. Ten magenta streamlines, equally distributed over a circumference of radius equal to $0.1D$, depart from $x = x_{LB}^N$.

of the reverse flow. The azimuthal velocity component distribution is asymmetric.

The last couple of images (figure 4.8e and figure 4.8f) clearly shows two vortical structures: the precessing vortex core and the outer secondary vortex. These two counter-winding co-rotating helices turn around the jet axis. The dissection of the snapshot (figure 4.8f) shows a significant recirculation region which penetrates the flow upstream to the nozzle.

4.1.3 Proper Orthogonal Decomposition analysis

The unsteady three-dimensional dynamics discussed above shows different vortical patterns for $S = 0$ and $S \neq 0$. A modal decomposition technique like POD (Berkooz et al., 1993) is suitable to identify the main features of the two flow fields (Lumley, 1967).

A no swirling jet is characterised by annular vortical structures which are clearly distinguishable farther than $2D$ from the nozzle (Violato and Scarano, 2011). Hence, the investigation volume extension along x direction is not sufficient to properly resolve such vortical structures. Hence, the application of Proper Orthogonal Decomposition for the no swirling case is ineffective. On the other hand, the typical structures of a swirling jet are noticeable since from the nozzle exit. Hence, chosen the swirl number $S = 0.61$, the three-dimensional main features of a free swirling jet are discussed in the following.

In figure 4.9 the energetic contribution of POD modes versus the mode number is reported. The inset shows the cumulative sum of the energetic contributions. Figures 4.10 and 4.11 report the Q isosurfaces of the first four modes of a swirling jet. They contain about 50% of the total energy included in the snapshots sampled. The isosurfaces presented are pictured two by two with high contrast colours for the sake of clarity. In each frame the red isosurface is related to the most energetic mode of the couple of modes pictured.

The first two POD modes, reported in figure 4.10a in red and blue colour, show two counter-winding spiral-shaped isosurfaces . They are located around the core jet region. Such structures are the most energetic ones of the whole flow field including 40% of the energy contained in the snapshots sampled. The whole vortical structure does not extends out

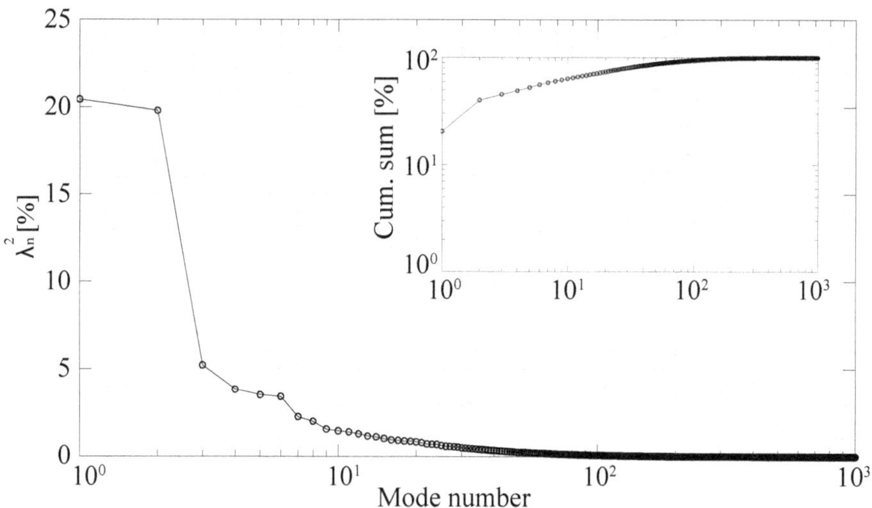

Figure 4.9: Energetic contribution of the POD modes versus the mode number in the velocity fluctuations basis for swirl number $S = 0.61$. The inset shows the cumulative sum of the energetic contributions.

of the nozzle. As a matter, it is representative of the precessing vortex (PV) described by Cala et al. (2006). The precessing frequency of such a structure can be deduced in two different ways: from snapshots inspection within the accuracy of the acquisition time resolution, or from the Fourier transform of the first two temporal coefficients vectors extracted from the Singular Value Decomposition (SVD) operated to obtain POD modes. The resulting Strouhal number of the PV precessing motion is $St = 1.08$ which is in accordance with literature (Nicholas Syred, 2006). Unfortunately, the limited range of investigated swirl ratios does not allow to identify a variation of Strouhal number with swirl number.

The other two modes presented in this section are reported in figure 4.10b. They are representative of 10% of the total energy included in the sampled snapshots. The spiral-shaped vortical structures involved are of two different kinds: the first one is totally included in the nozzle extension and wraps around a central vortical column in clockwise sense, the second one is located completely out of the nozzle extension. The cylindrical vortical structures are related to the central recirculation region.

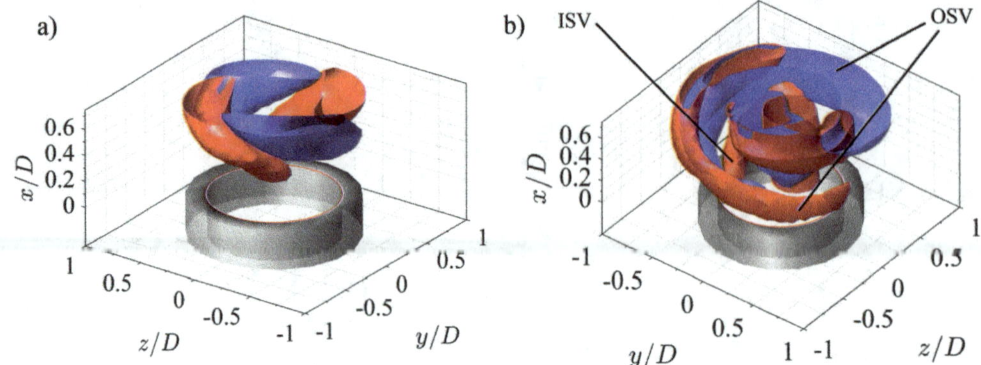

Figure 4.10: Q isosurfaces of the first four modes (50% of the total energy of the flow field) taken two by two; swirl number $S = 0.61$. a) 1st and 2nd POD modes (in red and blue colour respectively): 40% of total energy; b) 3rd and 4th POD modes (in red and blue colour respectively): 10% of total energy.

In conclusion, the first kind of structures is related to the inner secondary vortex (ISV) while the second one is related to the outer secondary vortex (OSV) (Cala et al., 2006).

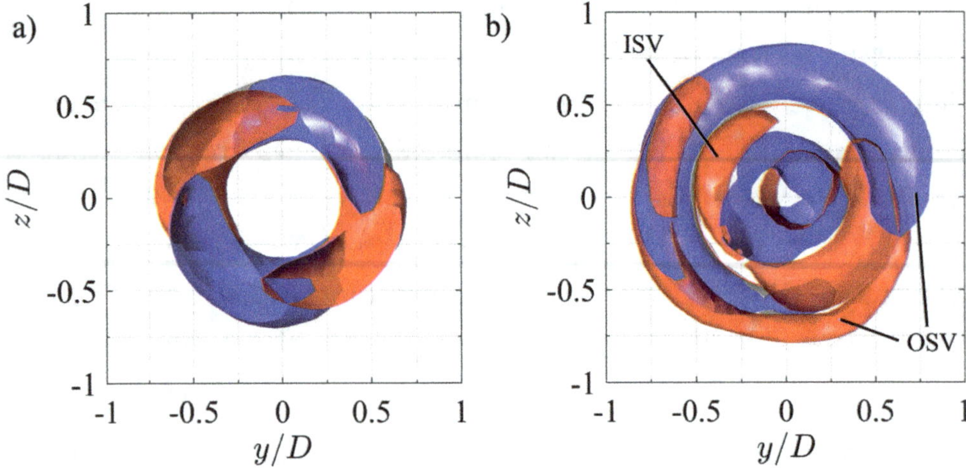

Figure 4.11: Top views of frames reported in figure 4.10. Q isosurfaces of the first four modes (50% of the total energy of the flow field) taken two by two; swirl number $S = 0.61$. a) 1st and 2nd POD modes (in red and blue colour respectively): 40% of total energy included; b) 3rd and 4th POD modes (in red and blue colour respectively): 10% of total energy included.

4.2 Flow field of an impinging swirling jet

In the present section the flow field related to an impinging swirling jet is presented. Among the nine designed swirl generators, five of them will be discussed in the following: $S = 0, 0.23, 0.43, 0.61, 0.74$.

The reference coordinate system is that reported in figure 3.10.

4.2.1 Close-to-wall velocity radial distributions

As done in section 4.1.1 the time-averaged velocity radial distributions will be presented. Then the discussion will focus on the Reynolds stresses.

The first case we will discuss in the following is the no-swirl case which is taken as reference case (figure 4.12). It is noticeable the significant decrease in the axial velocity with respect to the free configuration. This is caused by the approaching of the jet to the wall. It is known in literature that the influence of the wall on the flow extends $1D$ upstream along the axial direction for continuous impinging jets (O'Donovan, 2005). As a matter, the wall presence causes an adverse pressure gradient leading to an axial velocity decrease. In addition, the axial component of the no-swirl impinging case shows a local minimum on the jet axis and a maximum at $r/D = 0.4$. The double peak distribution is ascribed to the adverse pressure gradient which is generated by the impinging plate presence. This is confirmed by numerical predictions of Rohlfs et al. (2012) which were experimentally validated by Greco et al. (2016). After a sudden decrease of velocity in the range $0.5 < r/D < 1$, the axial velocity becomes basically negligible. This brings to the conclusion that the jet spreading affects the flow field close to the wall until $1D$ from the impingement.

The radial component of velocity is characterised by a maximum located at $0.7D$. The steepness of velocity distribution between the origin and this station is ascribed to the rapid deflection of velocity vector from the axial to the radial direction and the consequent contraction of streamlines where the wall jet is beginning to develop. Away from this station

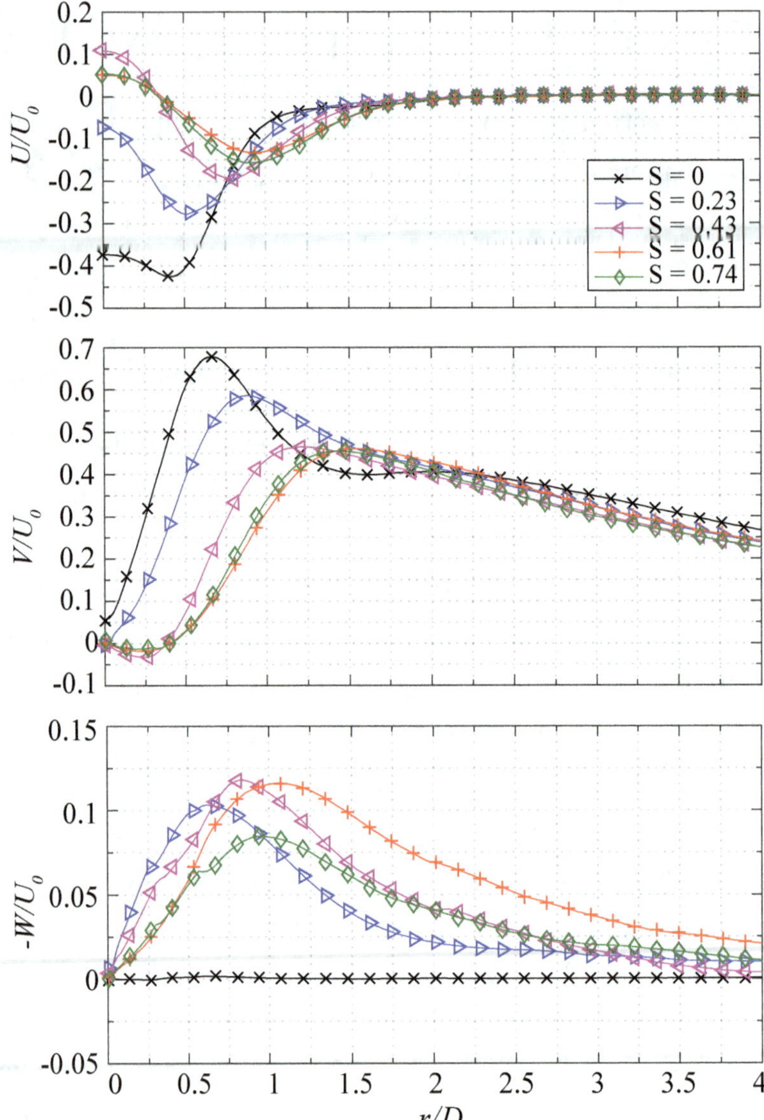

Figure 4.12: Radial time-averaged velocity distributions of an impinging submerged swirling jet ($x/D = 0.02$). Legend reports the swirl numbers.

the radial velocity experiences an opposite acceleration because of the increasing of the wall jet height. Once reached the distance of $2D$, the radial velocity linearly decreases with slope $-0.13 U_0/D$.

The first non-zero swirling case (i.e. $S = 0.23$) shows several differences from the impinging no swirling jet. The axial velocity distribution is significantly issued by a defect along the jet axis. In addition, the local maximum located at $0.5D$ is slightly displaced towards to the origin. The wider velocity distribution is symptomatic of the increased entrainment provided by the swirling jet. Hence, where the axial velocity is basically negligible in the no-swirl case, now is close to 10% of the bulk velocity U_0. The axial velocity becomes negligible at $1.5D$ from the origin.

The radial component of velocity is significantly influenced by the rotating motion. As a matter, the maximum has a smaller absolute value and it is located at $1.2D$. In addition, the linear decaying of velocity is anticipated at $r/D = 1.5$ respect to zero swirling case.

The radial distribution of the negative tangential velocity component is characterised by a positive slope $(0.26U_0/D)$ starting from the core. The distribution reaches a maximum located at $0.5D$ and then, the slope reverses with the same steepness until $2D$ from the origin where the velocity decays more gradually. This behaviour can be explained focusing separately on the two zones located at each side of the peak. To the left side of the maximum there is located the vortex core. In this region, the inertia of the fluid is small and the flow field is continuously fed with rotating motion. It behaves like a forced vortex: it is characterized by the increasing of the tangential velocity with the distance from the origin. On the other side of the peak, the velocity distribution is represented by fluid which exhibits a decaying swirling behaviour while flowing away from the axis. Hence, it behaves like a free vortex and is characterized by a hyperbolic decrease of the tangential velocity. The nearest analytical representation of this distribution is the Rankine vortex which is a typically adopted model of vortex in viscous fluids (Giaiotti and Stel, 2006).

$$w_{\text{vort}}(r) = \begin{cases} \Gamma r/\left(2\pi R^2\right) & r \le R \\ \Gamma/\left(2\pi r\right) & r > R \end{cases} \tag{4.2}$$

The increasing of the swirl number stresses all characteristics described above. Regarding to the $S = 0.43$ case, the axial velocity distribution is affected by a significant reverse flow in the range $0 < r/D < 0.4$. The maximum of reverse velocity flow is reached along the jet axis. Moving away from the origin, the axial velocity increases until $0.75D$ where is located a maximum of absolute axial velocity. The spread of the jet is more evident in this case where the axial velocity component is significant until $1.75D$.

The radial velocity distribution is affected in three different ways by the increasing of the swirl ratio: the occurrence of a reverse flow, the maximum radial velocity location is displaced further away from the origin, the maximum absolute value is significantly reduced. The first phenomenon is related to the occurrence of a recirculation in proximity of the impingement region. This makes new fluid to move to the impact area where the recirculation is located. The second phenomenon is symptomatic of the increased spreading of the jet before approaching to the wall. The resulting effect is that the onset of the spiral wall jet is postponed because the jet core gets wider. The overall spreading of the jet directly affects the velocity distribution which, in turn, experiences the lowering of the absolute value of the maximum.

Not surprisingly, the tangential velocity distribution increases its absolute maximum value with respect to the low swirl case. In addition, the spreading of the jet is still noticeable in this case because the absolute maximum location is displaced farther from the origin.

An overall interpretation of the flow field established at $S = 0.43$ can be provided gathering all the information given by the three velocity distributions. In proximity of the impingement area the flow experiences three different simultaneous motions: the fluid near the stagnation region, adjacent to the wall, is raised, new fluid is brought from regions out of the core to the impact area and a forced rotating motion generates a rotational flow field which envelops the core and induces the rotation away from it. The resulting motion is a spiral recirculation region extended in $0 < r/D < 0.5$.

The behaviour of the remaining swirl numbers investigated is not trivial. The axial velocity is still affected by a defect along the jet axis but this time the absolute value is lower respect to the previous swirling case. Hence, the reverse flow is still present in this case even if it is weaker than the previous at $S = 0.43$. The maximum, previously placed at $0.75D$, is now slightly displaced further away from the origin and it is characterised by a smaller absolute value.

The radial velocity component distributions for high swirl numbers are characterised by the asymptotic approaching of the maximum at $r/D = 1.5$, then the $-0.13U_0/D$ decay slope is followed. As a consequence the radial velocity gradients are smoother.

Unexpectedly, the tangential velocity component reverts its trend passing from $S = 0.61$ to $S = 0.74$. The lower swirl number is characterised by the displacement of the maximum further away from the origin (at $r/D = 1.1$). The maximum is approximately equal in value to the one related to $S = 0.43$. On the other hand, the tangential velocity component distribution related to $S = 0.74$ experiences two unexpected phenomena: the maximum gets closer to the origin and its absolute value significantly decreases. This behaviour can be interpreted as the result of the engulfment of the fluid. A rotating motion is still imparted to the fluid but in this case not all angular momentum imparted to the fluid at the nozzle is efficiently transmitted until to the wall.

Figure 4.13 reports the radial distributions of the Reynolds stresses in a plane placed at $x/D = 0.02$ from the wall.

In the no-swirl case the $\overline{u'^2}$ radial distribution is characterised by four distinctive peculiarities: a minimum located in correspondence of the jet axis, two local maxima located at $0.625D$ and at $2D$ from the origin and a linear decaying with $-4 \cdot 10^{-3}U_0/D$ slope from $r/D = 2.125$. The first critical point is caused by the jet potential core which is not still vanished and the fluctuations are small in proximity of the origin. As a matter, the shear layer embracing the core does not have enough space to merge with it in the distance between the nozzle and the wall. The inner peak could be ascribed to the shear layer impact to the wall. A slight spread of the jet

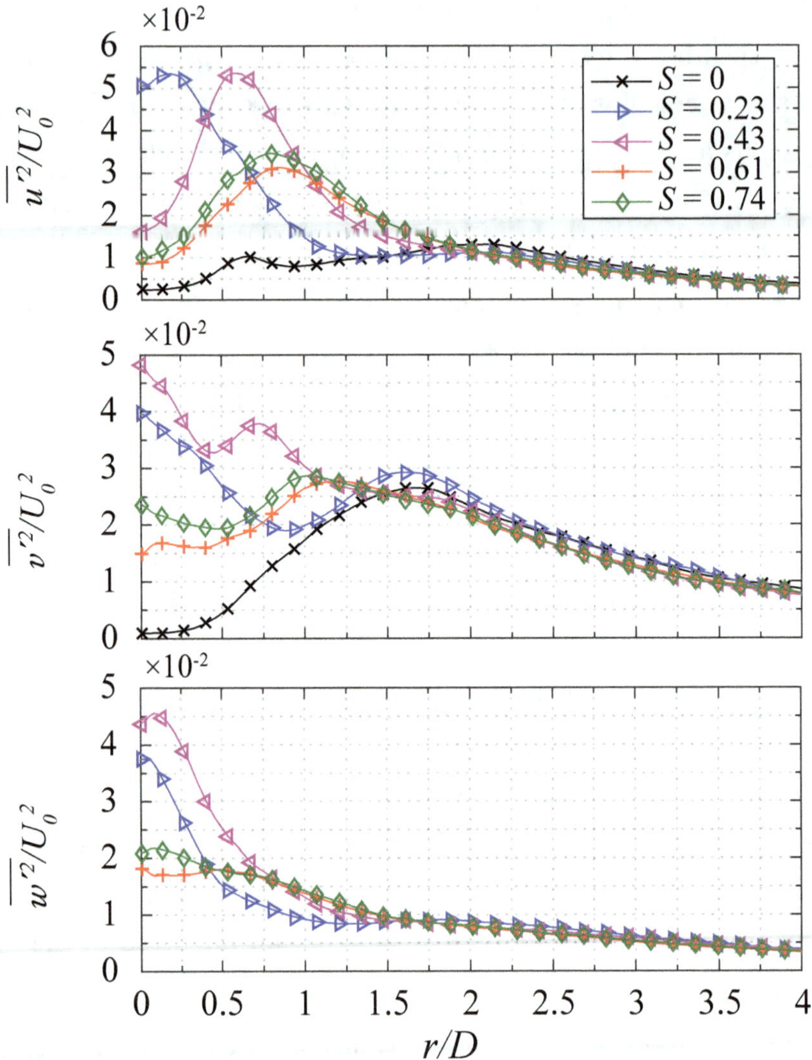

Figure 4.13: Radial distributions of Reynolds stresses for an impinging submerged swirling jet ($x/D = 0.02$). Legend reports the swirl numbers.

justifies the displacement of the critical point from $0.5D$. Also an outer peak is observed at $2.25D$ from the origin which can be ascribed to the occurrence of a secondary vortex ring (Greco et al., 2016). The positions of the axial peaks correspond to those found by O'Donovan (2005) for impinging continuous jets.

The second-order moment of radial component of velocity shows a

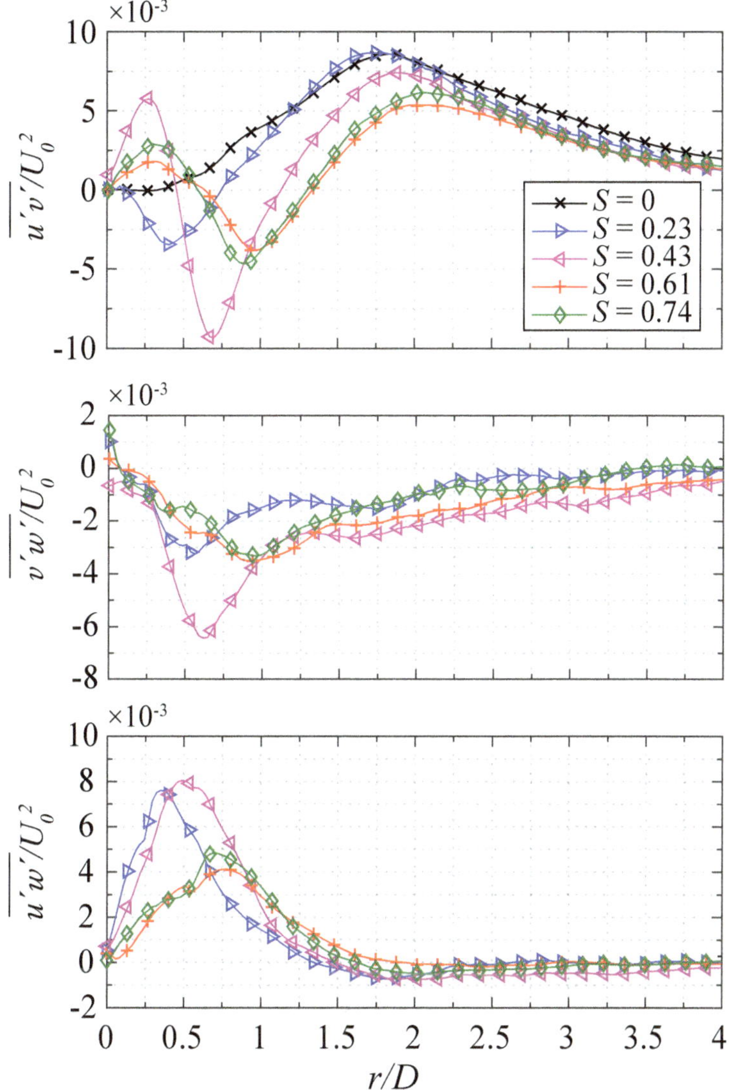

Figure 4.13 (Cont.): Radial distributions of Reynolds stresses for an impinging submerged swirling jet ($x/D = 0.02$). Legend reports the swirl numbers.

plateau which extends from the jet axis to $0.25D$. Moving away from the origin the quantity increases and reaches its highest value at $r/D = 1.625$. After that station follows the decay with $-8 \cdot 10^{-3} U_0/D$ slope.

The only non-zero mixed second-order moment for the no swirling case is the $\overline{u'v'}$ which resembles the $\overline{v'^2}$ distribution. In fact, the plateau in proximity of the impact and the maximum located at $r/D = 1.75$ are still present. The decaying slope after the maximum station is $-2.5 \cdot 10^{-3} U_0/D$.

The introduction of a rotating motion ($S = 0.23$) causes a significant change in the radial distributions of second-order moments. The $\overline{u'^2}$ experiences a minimum along the jet axis and a maximum located at $0.25D$. After that station, there is a sudden decrease. After $r/D = 1.5$, the quantity keeps low in value experiencing a linear decay from the radial distance equal to $r/D = 2.125$. The high energetic level of turbulence along the x axis can be ascribed to the core of the jet which is breaking before impacting to the wall. As a matter, even if the swirling rate is low this still causes instabilities in the conical shear layer and a premature break up of the core of the jet.

The turbulent content of radial component of velocity along the jet axis is high too. Moving away from the origin, the negative sloping curve reaches a minimum located at $0.875D$, then a maximum is reached at $1.625D$. After that station, the same turbulence decay encountered in other cases occurs. The high energetic content along the radial direction in the range $0 < r/D < 0.875$ is symptomatic of two effects caused by the low swirl ratio: anticipated shear layers merging and noticeable spreading of the jet. The first effect can be deduced by the high values of turbulence content while the second phenomenon clearly appears from the widening of the peak prominence of the first peak. The second peak presence suggests the similarity with the no-swirl case.

The high turbulence level in the nearby of the impingement region occurs also in the $\overline{w'^2}$ radial distribution. It is characterised by a maximum along the jet axis and then a gradual decay. As noted before, the high turbulence level is ascribed to the anticipated shear layer merging.

The behaviour of $\overline{u'v'}$ still has some similarities with the no swirling case: a maximum located at $1.75D$ and the linear decay after that station. In addition, there is another minimum located at $0.375D$. This oscillating trend shows the variable contributions of axial and radial turbulence along the radial distribution of the Reynolds stress.

The $\overline{v'w'}$ shows a maximum in absolute value located at $0.5D$ and then a slow decay moving away from the origin.

The last Reynolds stress investigated is $\overline{u'w'}$ which resembles to some extent the behaviour of $\overline{u'^2}$. A maximum is located at $0.875D$ then a rapid decay happens. After the $1.25D$ station, a change in sign appears.

Increasing the swirl number significantly affects the Reynolds stresses' distributions. The first Reynolds stress $\overline{u'^2}$ is characterised by the decrease in absolute value of the local minimum located along the jet axis and by the shift in the maximum location from $0.25D$ to $0.6D$. The first effect can be ascribed to the core of the jet impacting on the wall. This results in a low turbulence region in proximity of the jet axis. In this case the swirling motion seems to keep solid the inner region of the flow until it reaches the impinged surface. The turbulent energetic content of the axial component of the flow suddenly increases reaching a maximum which, with respect to the low swirling case ($S = 0.23$), is not affected on the absolute value by the increased swirl ratio. The second effect noticed, the shift in the maximum location noticed comparing the $\overline{u'^2}$ radial distributions for two different swirling ratios ($S = 0.23$ and $S = 0.43$), can be ascribed to the spreading of the jet which is significant in this case. Increasing the swirl number further than $S = 0.43$ causes the lowering of the absolute maximum value and an additional shift away from the origin. The maximum is now located at $0.8D$. After the maximum, all the moderate to high swirl ratio distributions share a common decaying trend.

The radial distributions of $\overline{v'^2}$ for moderate to high swirl numbers are characterised by high turbulence along the jet axis, a minimum and a maximum. In particular, the $S = 0.43$ distribution shows, with respect to $S = 0.23$, the increasing in the maximum along the jet axis, the approaching of both, the minimum and the maximum, towards the centre of the jet. The new locations of the critical points are $0.375D$ and $0.75D$ respectively. This results in the fact that, even the mixing region is wider, the main part of the radial component of turbulence is included in a circle of radius equal to $1D$. This station corresponds to the turbulence decay. It is worth noting that the increasing of swirl number from low ($S = 0.23$) to moderate ($S = 0.43$) values corresponds to an increase of the turbulent content of all radial Reynolds distributions. Further increasing of the swirl number over $S = 0.43$ causes the turbulence decreasing along the jet axis. The minimum and the maximum patterns are still noticeable but they significantly decrease in absolute value with respect to the previous swirl ratio. As a matter, passing from moderate to high swirl numbers ($S \geq 0.5$) implies a general lowering of the whole distribution.

The Reynolds stress distribution related to tangential component of velocity, $\overline{w'^2}$, for a moderate swirl number ($S = 0.43$) shows a distribution which is entirely located over the $S = 0.23$ curve. Increasing the swirl number cause, also in this case a general lowering of the whole distribution.

The mixed second-order moments' distributions confirm the behaviour previously described: the switch from low to moderate swirl numbers results in a general increase of turbulent content of flow field. The mixed Reynolds stress $\overline{u'v'}$ distribution is characterised by a significant increase of the minimum and of the maximum absolute values. A shift of both critical points towards the origin is also noticeable. The behaviour of moderate and high swirl numbers is coherent with the other distributions, thus a general reduction in absolute value of all critical points is verifiable.

The remaining mixed Reynolds stresses investigated $\overline{v'w'}$ and $\overline{u'w'}$ experience stronger gradients in proximity of critical points switching from low to moderate swirl number, while it is noticeable an overall lowering of all recognisable patterns passing from moderate to high swirl numbers. In addition, at high swirl numbers, all critical points experience a slight shift further away from the origin.

4.2.2 Adiabatic wall temperature in incompressible flows

In the following, the radial distributions of the time-average and standard deviation of wall temperature will be presented. In order to properly scale the first quantity, a non-dimensional parameter has to be introduced.

The assumption according to which the ambient temperature where the jet exhausts is constant during the test is not correct. Hence, the general problem involved in the present experimental apparatus is to properly non-dimensionalise the wall temperature distribution thus obtaining a parameter, called η, that takes into account the influence of the swirl number on the ambient temperature. The scaling adopted for the wall temperature distribution is reported in the following expression:

$$\eta = \frac{T - T_\infty}{T_j - T_\infty} \tag{4.3}$$

where T is the time-average temperature distribution, the jet temperature T_j is assumed to be constant for all swirl numbers and T_∞ is the ambient temperature in which the jet exhausts. This last parameter depends on the swirl number. In order to estimate T_∞, for each test, the temperature in proximity of the impingement area is acquired at four critical time steps: at the beginning of the test (T_{Init}) as check of the uniformity of temperature through the whole tank, at the beginning of the thermal acquisition ($T_{\text{InitTherm}}$), at the beginning of the PIV acquisition (T_{InitPIV})

Swirl number	$T_{\textbf{EndTherm}}$ $[^\circ\text{ C}]$
0	20.50
0.23	20.60
0.43	20.65
0.61	20.70
0.74	20.80

Table 4.2: Correspondence between the swirl number and the temperature in proximity of the impingement at the end of the thermal acquisition.

and at the end of the thermal acquisition (T_{EndTherm}). The last one will be assumed as the temperature which best approximates the ambient temperature T_∞ in which the jet exhausts (table 4.2).

4.2.3 Wall temperature radial distributions

The time-averaged and the normalised standard deviation temperature radial distributions are presented in figure 4.14.

The no-swirl case is characterised by a plateau until $r/D = 1.5$ and a slow decrease of the time-averaged wall temperature moving away from the impact centre. The behaviour of the time-averaged wall temperature is strictly correlated to the standard deviation of temperature σ. This has a large plateau which extends to $1D$, a steep increase and a maximum located at $2.25D$. After that station the temperature fluctuations are progressively smaller. As it can be noticed, the two plateaus corresponds one to each other. In particular, until the fluctuation of temperature has not reached its maximum, η is basically equal to unit. This suggests to think that the station until σ is small spots the width of the core jet which thus basically reaches undisturbed the wall. The maximum of σ indicates the secondary vortex typical of an impinging conventional jet. Hence, both curves confirm that the circular jet case basically travels undisturbed toward to the wall because it has no room to completely develop.

A low swirl number involves significant differences from the previous discussed case. The first effect is a lowering of the maximum related to σ distribution. In addition to this, there exists a general increasing in temperature fluctuation in the range included between the centre of impact and the maximum location. Hence, the addition of an even small tangential motion to a circular jet results in a general higher turbulence level at the wall. The augmentation of σ is particularly significant at $1D$. This can be ascribed to the shear layer instabilities which the rotating motion involves resulting in enhanced mixing between jet and surrounding fluid at rest, even in the small distance between the nozzle and the impinged

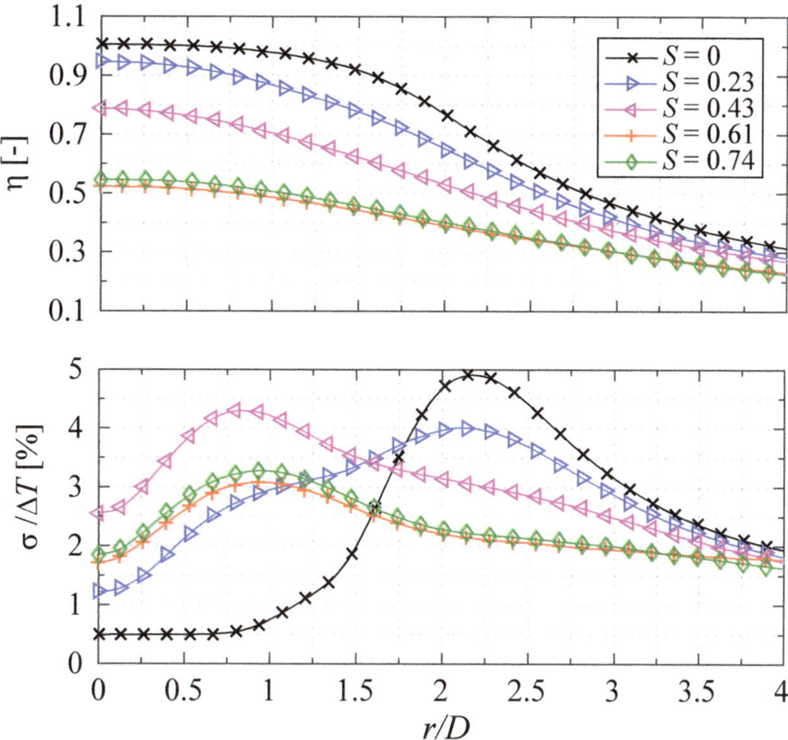

Figure 4.14: Wall radial distributions of non-dimensionalised time-average η and standard deviation σ of wall temperature for an impinging submerged swirling jet. Legend reports the swirl numbers.

wall. The radial distribution of time-averaged wall temperature differs from the conventional case for the lack of the plateau and for a general lowering of the quantity distribution. The first information, which was previously useful to spot the end of the potential core jet, now gives a dual information: the core jet is disturbed by the swirl motion and the boundary between the core jet and the outer region from the shear layer is not clear as before. The effect of the general lowering of η distribution confirms the mixing enhancement provided by the addition of even a small swirling motion to a circular jet.

Increasing the swirl number from low ($S = 0.23$) to moderate swirl number ($S = 0.43$) brings additional mixing all over the range of investigation. This can be deduced from a further general lowering of the η distribution. The σ distribution for $S = 0.43$ basically differs from the low swirl number ($S = 0.23$) case for the maximum location which, previously located at $2.25D$, moves in the nearby of the impact centre placing at $0.75D$. The maximum level of temperature fluctuation along the jet axis increases further in this case.

Further increasing of the swirl number to high values ($S = 0.61-0.74$) provides a common asymptotic behaviour. The wall temperature seems to reach an asymptote and the two η distributions barely differ one from each other. As a consequence, the two swirl numbers behave as the same from the time-averaged temperature distribution and mixing points of view. Focusing on the shape and magnitude of σ, the two distributions basically overlap experiencing a general lowering. This can be ascribed to the anticipated breaking of the vortex core. As a consequence, the jet flow has space to better mix with the ambient fluid before reaching the impinged wall.

4.2.4   Temperature time sequences

In order to simultaneously extract spatial and temporal information about the organisation of the unsteady wall thermal field of an impinging swirling jet, the first half of the synchronised thermal acquisition is taken into account. Within this time frame, the temperature fluctuations T' time sequences for five swirl numbers are extracted from three circumferences centred on the origin having radii respectively equal to $0.8D$, $1.5D$ and $2.3D$. Such radii are chosen in order to describe three different representative behaviours.

The first investigated time sequence (figure 4.15) shows a significant difference depending on the swirl number. The no swirling case is characterised by small temperature fluctuations except for some blue spots. Such a behaviour can be explained referring to figures 4.12 and 4.14. As it can be noticed, the distributions of U/U_0 and σ from the origin to $r/D = 0.8$ show respectively high values of axial velocity component and small standard deviation values of temperature fluctuation. This means that, at this location, we are investigating the impingement region within

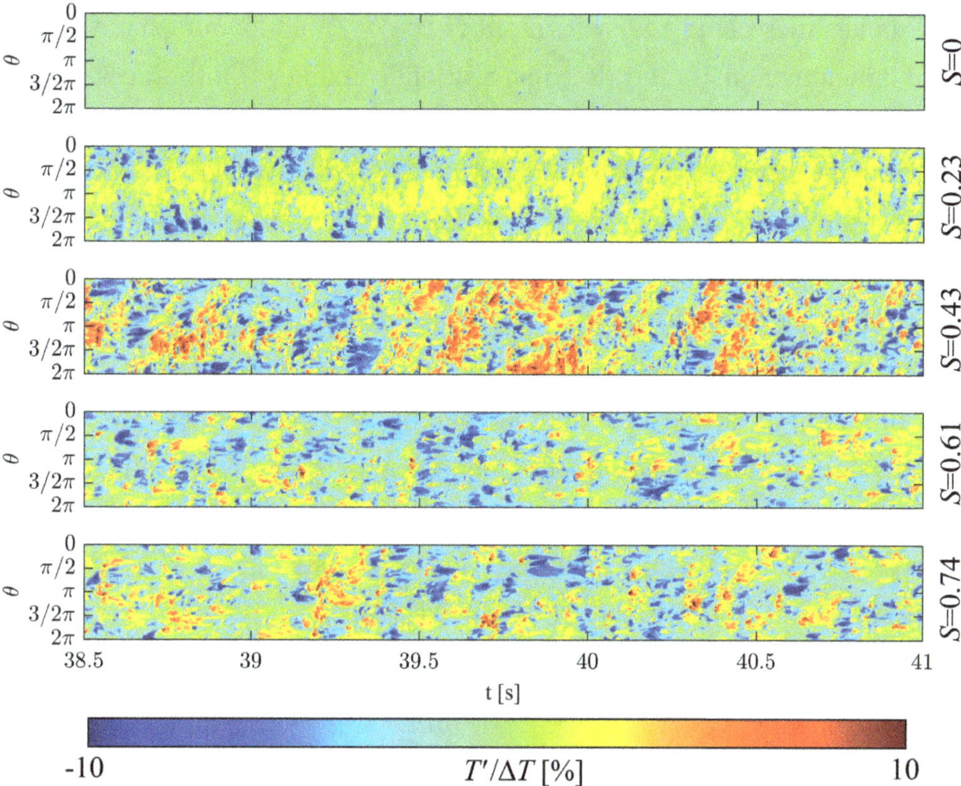

Figure 4.15: Time sequences of temperature fluctuations at $0.8D$ from the origin. The duration of each time sequence is equal to 2.5 seconds starting from the beginning of the synchronised acquisition. Swirl numbers are reported on the right side of each stripe. The scale on the left side reports the azimuthal coordinate in radians. On the bottom, the time scale is reported. The colorbar refers to the temperature fluctuation percentage respect to ΔT imposed.

the core jet. Hence, the flow does not mix at all with the ambient at this stage.

The superimposition of a tangential motion triggers the entrainment phenomenon since short distances from the impact. The general behaviour shows large positive fluctuation regions together with several strongly negative spots. Such a pattern can be ascribed to the occurrence of both enhanced entrainment phenomenon and mixing performances of the swirling flow field.

Further increasing the swirl number to $S = 0.43$ favours the entrainment phenomenon to the detriment of the mixing performances. As a matter, the time distribution is disseminated of well defined positive (in red) and negative (in blue) high temperature fluctuations spots. This means that the swirling motion is acting to strongly drag ambient fluid toward the impingement centre but it is not able to mix enough the fluids at different temperatures.

The mixing performances suddenly improve for $S = 0.61$. The dimension of cold and hot regions is much smaller respect with the previously discussed case. The entrainment phenomenon brings new cold fluid towards the jet core while the mixing is dramatically increased. This is in accordance with the significant lowering of η distribution at $r/D = 0.8$ passing from $S = 0.43$ to $S = 0.61$ (figure 4.14).

The last swirl number investigated for such a distance from the origin ($S = 0.74$) demonstrates worse mixing performances. As a matter, red spots are large again while the entrainment appears to drag more cold fluid towards the impingement region.

In order to give a complete framework of the temperature fluctuations dynamics, another distance from the origin (i.e. $r/D = 1.5$) is investigated (figure 4.16).

Moving to such a station, the no swirl case shows a significant increasing of the entrainment contribution. As a matter, the related time-distribution is characterised by several and finely distributed negative fluctuation spots. On the other hand, the most frequent value of temperature fluctuations is close to zero. This means that the core jet has still strong influence at this station.

The time-distribution related to the first swirling case ($S = 0.23$) shows a significant difference from $r/D = 0.8$: there are more cold and hot spots. Their simultaneous presence suggests that the mixing performances are affected negatively moving away from the origin. Another significant difference is the blue spots extension along the time dimension. This can be explained as the occurrence of radial velocity compo-

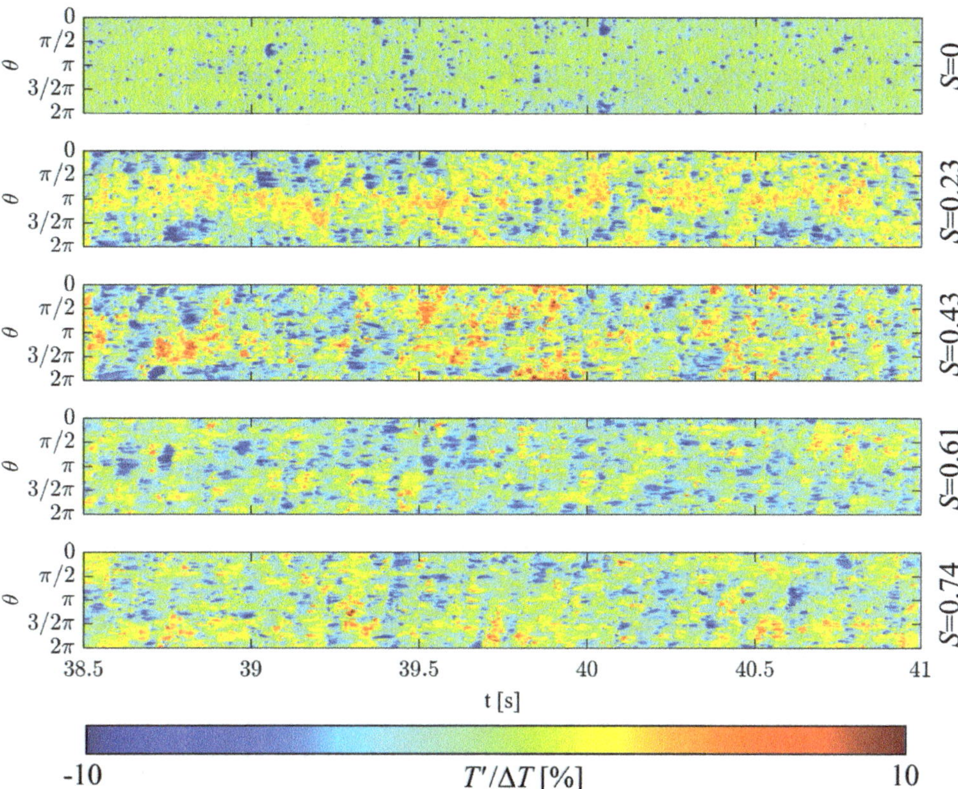

Figure 4.16: Time sequences of temperature fluctuations at $1.5D$ from the origin. The duration of each time sequence is equal to 2.5 seconds starting from the beginning of the synchronised acquisition. Swirl numbers are reported on the right side of each stripe. The scale on the left side reports the azimuthal coordinate in radians. On the bottom, the time scale is reported. The colorbar refers to the temperature fluctuation percentage respect to ΔT imposed.

nent acceleration between $r/D = 0.8$ and $r/D = 1.5$ which results in the enhanced entrainment of ambient fluid.

The time-distribution related to swirl number $S = 0.43$ shows several hot spots as in the previous case. The main difference from before is in their dimensions which is dramatically reduced. This means that at this distance from impingement region and with such a swirl number, a good equilibrium between the entrainment and the mixing phenomena is reached.

Further increasing the swirl number to $S = 0.61$ affects the mixing in a positive way at $r/D = 1.5$. In fact, red spots are barely visible at this distance. The entrainment effect is still noticeable because of the presence of blue regions but of smaller dimensions. This is symptomatic of the increased level of turbulence at further stations from the impact (figure 4.14).

The reduction of cold and hot spots dimensions is noticeable also for $S = 0.74$. The higher level of turbulence at $r/D = 1.5$ favours the mixing resulting in a distribution much similar to $S = 0.61$. This suggests that the performances of the two flow fields are comparable at these conditions.

The last station investigated is at $2.3D$ from the origin (figure 4.17). The stripe related to swirl number equal to zero is characterised by uniformly distributed hot and cold spots. The main noticeable characteristic of such a stripe is the different size of spots. It appears to strongly depend on whether the spot is related to a high or low value of fluctuation. The hot spots, related to the jet, feed the flow field with hot water. The cold ones are related to the entrainment phenomenon which brings cold fluid from the external region of the jet. As a matter, the hot spots are smaller than the cold ones. This difference can be explained supposing that the impinging structures are well distinguishable one from each other also after the impact. This does not promote the equilibrium between entrainment and mixing. Hence, the lack of a good mixing provides spots perfectly distinguishable along θ direction and frequent variations of temperature fluctuation on the wall along the time direction.

The stripe related to $S = 0.23$ (low swirl number) reports both similarities and differences with the previous case. The noticeable difference in this case is that the spots with positive temperature fluctuation value are significantly larger and well spread all over the circumference while the cold spots appear as better distinguishable. In addition, they are surrounded by extended low fluctuation regions. This is symptomatic of the enhanced mixing behaviour of the flow field.

The further increasing of swirl number to a moderate value ($S = 0.43$)

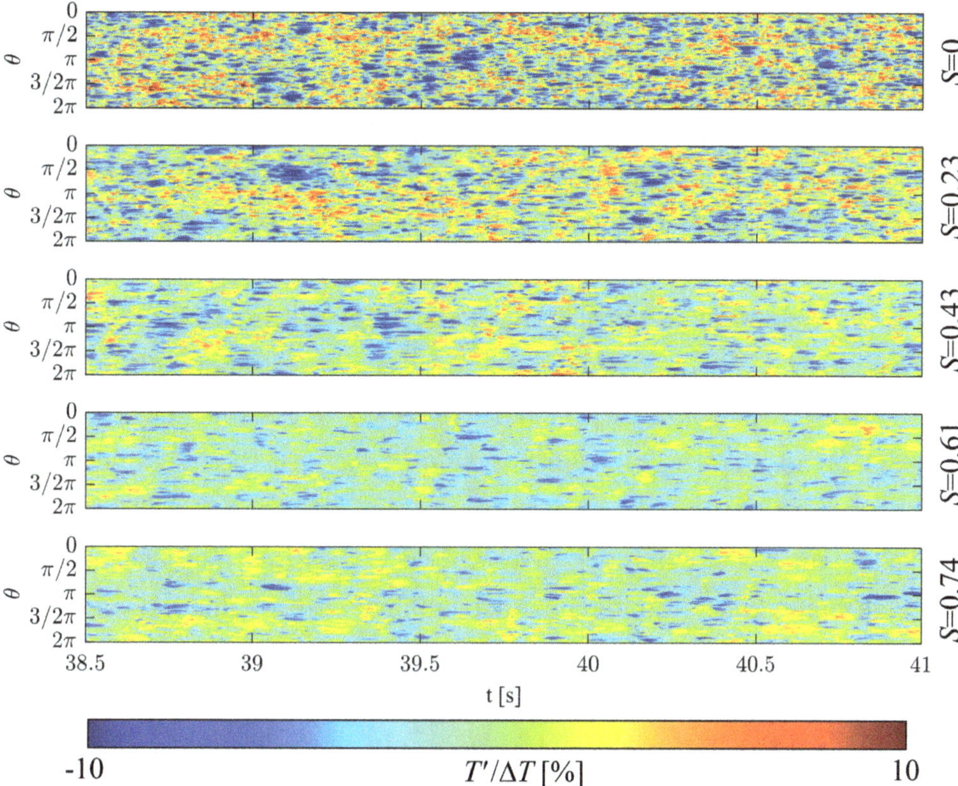

Figure 4.17: Time sequences of temperature fluctuations at $2.3D$ from the origin. The duration of each time sequence is equal to 2.5 seconds starting from the beginning of the synchronised acquisition. Swirl numbers are reported on the right side of each stripe. The scale on the left side reports the azimuthal coordinate in radians. On the bottom, the time scale is reported. The colorbar refers to the temperature fluctuation percentage respect to ΔT imposed.

generates a different behaviour in the widening of the jet. Along the circumferential direction, the spots have a similar extension with respect to the previous case. The significant difference in this case is that the tracking of spots of temperature fluctuations is more difficult. The mixing at $2.3D$ away from the impingement centre brings a significant lowering of temperature fluctuations. This could be ascribed to the higher turbulence level caused by the swirling flow field (figure 4.14). As a consequence, the fluid supplied by the jet mixes well with the cold one in proximity of the wall and, after $2D$, the temperature does not fluctuate much while the velocity fluctuations are still significant as shown in section 4.2.1.

The stripe related to the swirl number equal to $S = 0.61$ is characterised by an overall lowering of the absolute value of temperature fluctuations. This can be explained by the occurrence of higher turbulence levels in the velocity flow field. A simple interpretation of this is that the vortical structures which crawl along the wall are smaller because they completely breakdown after the impact. Furthermore, the positive temperature fluctuations magnitude are barely noticeable. On the other hand, the cold areas still act an important role with their significant extension along the time direction.

The highest swirl number investigated $(S = 0.74)$ shows a counter intuitive behaviour. Taking into account the size of fluctuation spots, the frequency of occurrence of high and low fluctuation spots and the mean value of the temperature fluctuation, the overall sensation is that the present case is a mid-way between $S = 0.43$ and $S = 0.61$ cases. The red spots are noticeable again. This suggests that there exist local increments in the temperature fluctuations. Hence, the mixing behaviour is worse than in the previous case.

In conclusion, the unsteady temperature fluctuation maps related to swirl number $S = 0.61$ appear to fruitfully join the effects of both entrainment and mixing phenomena. As a matter, the mutual cooperation of these phenomena produces a favourable mixing environment.

4.2.5 Wall velocity radial distribution

As described by Warhaft (2000), small temperature differences have no dynamical effect on the fluid motion itself. For this reason, the temperature differences can be used as an additional tracer in the flow under investigation. The time sequences presented in section 4.2.4 are characterised by high and low temperature fluctuation spots with a significant signal to noise ratio. This suggests the possibility to apply a cross-correlation algorithm which involves the MGIWD approach (Discetti and Astarita,

2010) to evaluate the displacement of the patterns. This results in estimating the velocity at very short distance from the wall which could differ in some way from the one estimated with the application of the T-PIV technique.

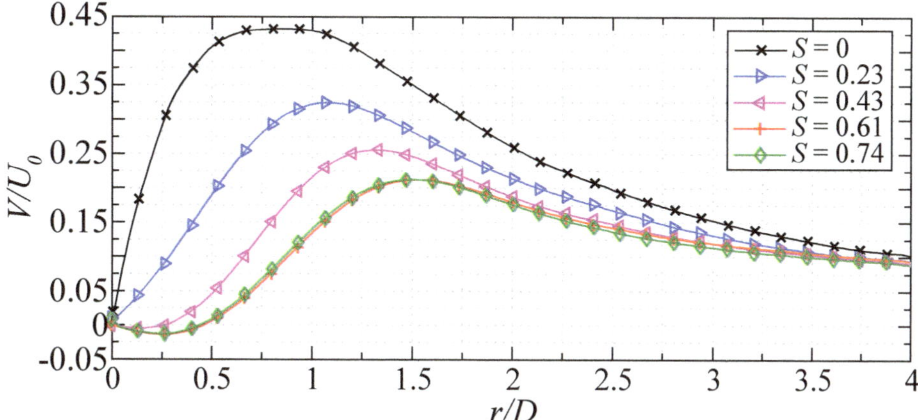

Figure 4.18: Time-averaged radial velocity component distributions estimated from the temperature fluctuation images.

The cross-correlation data processing procedure is the same of which adopted in a classic planar PIV algorithm. Hence, the thermal images are processed with final square windows equal to 16×16 with overlap set to 75%. In order to validate data, the outlier detection method proposed by Westerweel and Scarano (2005) is employed.

Figure 4.18 reports the time-averaged distributions of radial velocity component at five swirl numbers obtained from a sequence of 16 000 images. The no swirl case is characterised by a steep increasing of velocity in proximity of the impingement centre. A plateau is located in the range $r/D \in [0.5, 1]$ where the velocity reaches a maximum. After $1D$ from the origin the velocity gradually decreases.

The introduction of a low swirling motion causes the displacement of the maximum location away from the impingement centre. In addition, the absolute value decreases. A moderate swirl number, i.e. $S = 0.43$, involves the occurrence of a weak recirculation region in the range $0 < r/D < 0.25$. The inverse motion toward the impingement cen-

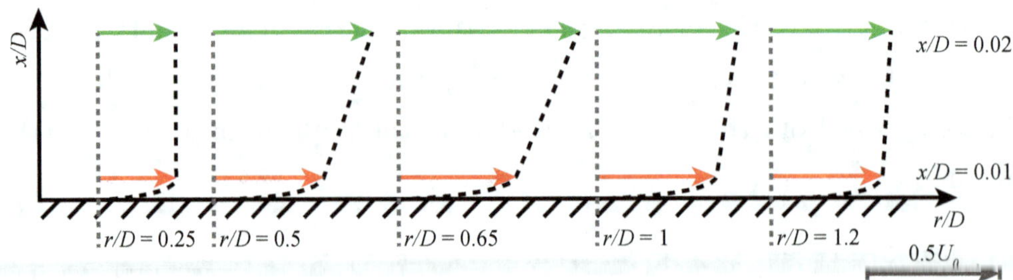

Figure 4.19: Comparison between radial velocity component at different locations along the wall according to PIV (arrows in green) and thermography data (arrows in red) for a conventional round jet ($S = 0$). The dimensions of arrows are in scale. Reference arrow reported in grey color.

tre can be ascribed to the presence of a localised vortical structure which acts engulfing the fluid in the nearby. Starting from $r/D = 0.25$, the velocity rapidly increases reaching a maximum at $1.25D$ away from the impingement centre. The absolute value of this peak is lower than the previous case.

The last two curves related to high swirl numbers ($S = 0.61$ and $S = 0.74$) almost overlap one to each other. They experience a further displacement of the maximum location away from the impingement centre and a general lowering of the velocity distribution. In addition, the recirculation zone in proximity of the impingement centre widens extending to $r/D = 0.5$.

The comparison of figure 4.18 with figure 4.12 gives additional information about the development of the flow field in the few fractions of diameter between the investigation region related to T-PIV volume and the wall velocity data above discussed. As a matter, the average radial distributions obtained with the two techniques appear to be in full accordance: the maximum locations perfectly correspond. As expected, due to the shear stress in the thermal boundary layer, figure 4.18 shows velocity values lower than those reported in figure 4.12. The main difference between them provides a further information about the flow field in case of the absence of tangential motion. As a matter, the velocity distribution measured with the temperature fluctuation distribution shows a plateau which is narrower in the T-PIV data. Hence, this suggests that the switching from impinging region to wall jet region follows the dynamics reported in the schematic in figure 4.19. In proximity of the impingement centre ($r/D = 0.25$) the velocity reported by the two distributions are equal. In each profile of time-averaged radial velocity component the black dashed line is the supposed linear matching of the information belonging to the two investigation heights. At $r/D = 0.5$ the V component of velocity measured by both techniques experiences a sudden acceleration. Proceeding away from the impingement centre the velocity on the last slice of T-PIV volume further increases while the velocity estimated from the thermal images is constant until $r/D = 1$. Such a location corresponds to the plateau in figure 4.18. On the other hand, the T-PIV velocity estimated at x_{LB}^{W}/D continues to increase until $r/D = 0.65$, then it suddenly decreases. After $1D$ from the impingement center the two velocity profiles come back to coincide one to the other. This means that the velocity gradients are smooth enough to slowly penetrate the last fraction of diameter along x direction and influence the thermal boundary layer.

4.2.6 Three-dimensional flow field

In the present section the unsteady three-dimensional flow field of an impinging swirling jet in proximity of the wall will be discussed. The reference system adopted and the measurement boundaries extracted from the illuminated volume are reported in section 3.4.

Figure 4.20 reports two snapshots of the instantaneous flow field of a no swirling jet impinging on a flat wall located at distance $H/D = 2$. The information reported in the figure are the following:

- isosurface corresponding to $Q = 0.3$ contoured with the azimuthal velocity component w;

- black streamlines placed at the boundaries of investigation volume (vertical planes located at $z = z_{HB}^{W}$ and at $y = y_{HB}^{W}$). The contoured background is related to w;

- grey isosurfaces related to $u/U_0 = 0.05$;

- red streamlines depart from a circumference of radius $0.25D$ located at x_{HB}^W;

- grey streamlines depart from a circumference of radius $0.75D$ located at x_{HB}^W;

- the plane located below the investigation domain is contoured with the instantaneous temperature fluctuations T' and it is positioned at $x = 0$.

Figure 4.20a reports that the overall organisation of the flow field can be distinguished in two possible occurrences: the crawling of annular vortices and their breakdown. The instantaneous flow field reported in figure 4.20a shows a large vortical structure which embraces the whole impingement region and several annular vortical structures centred on the impingement centre. In particular, the first vortical structure, located in proximity of the origin, can be ascribed to the impact of the shear layer with the wall. The second kind of vortical structures is characterised by ring-shaped patterns. These structures alternate one from another with a distance of $0.5D$. In addition, during their leaving from the origin, they do not break up for a long distance. As a matter, their shape is recognisable until $2D$ from the origin. After that station, instabilities arise and the annular vortices distort. After $r/D = 2.5$ the vortices break. As it can be noticed by the contour over the annular structures, the w component of velocity is close to zero until $2D$. Hence, the ring-shaped vortices are not affected by any significant azimuthal spatial gradients.

The absence of tangential motion is confirmed focusing the attention on the red and grey streamlines. They are basically aligned to x direction on the upper part of the observation domain. With the approach to the wall, the streamlines smoothly adapt spreading radially.

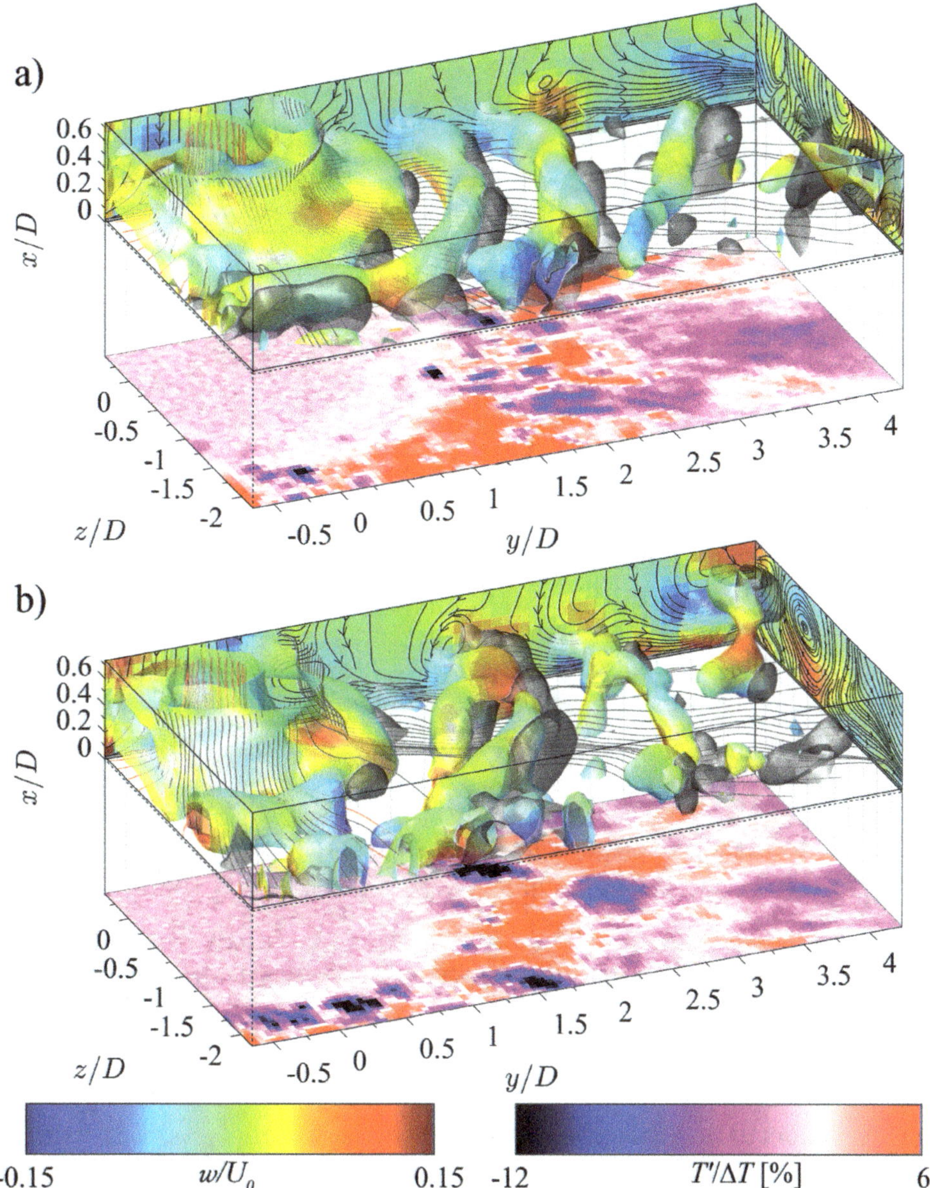

Figure 4.20: Three-dimensional representation of two instantaneous flow fields of the no-swirling jet ($S = 0$) suspended at $H/D = 2$ over a flat wall. The contoured isosurface corresponds to $Q = 0.3$. The left colorbar is related to azimuthal component of velocity. The grey isosurfaces are related to $u/U_0 = 0.05$. The second colorbar is related to temperature fluctuation. The red and grey streamlines start from two concentric circumferences located at $x = x_{HB}^W$ centred on the jet axis and of radii respectively equal to $0.2D$ and $0.75D$.

Figure 4.21: Schematic of the mutual interaction phenomenon between consecutive annular vortices.

In the snapshot reported, some positive u regions are noticeable (grey isosurfaces). They are located downstream of the ring-shaped vortices. These regions can be explained as the direct effect of the fluid drag from the wall because of the rotation of annular vortices. As it can be noticed by the black streamlines located at $z = z_W^{HB}$, the vorticity component along the z direction is negative. As a consequence, the vortex in proximity of the impingement centre takes the hot jet fluid and pushes it downward to the wall (figure 4.21). Once the hot fluid passes below the vortex, it comes up at lower temperature because of the mixing with the fluid at rest. Focusing on stations located farther from the origin, other two vortices are crawling. They are both smaller than the first one but act similarly. Moving away from the origin, the mixing phenomenon gradually weakens. As a result, each of the crawling vortices experiences a different flow temperature above them. This causes gradually lower temperature fluctuations measured on the wall at farther stations from the origin.

The discussion presented above about the snapshot reported in figure 4.20a shows the reasons of the bad mixing capabilities of the no swirling case. The high coherent organization of the crawling vortices does not help the mixing. As a matter, the temperature fluctuations map is characterised by a high non-uniform distribution which means that the fluid is often away from its time-averaged value.

Figure 4.20b reports the occurrence of the breakdown of the annular

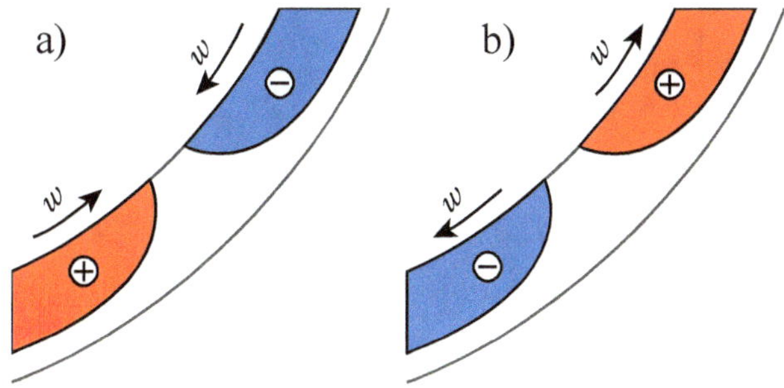

Figure 4.22: Schematic of the occurrence of spatial gradients of azimuthal component of velocity. a) Long life annular vortex. b) Short life annular vortex.

vortical structures. Particularly interesting is the presence of small spatial gradients of w velocity component. This can be ascribed as one of the phenomena concurring to vortices breakdown. In case of w gradients along the extension of a vortical annular structure, the relative positioning of positive and negative azimuthal velocity regions is of fundamental importance to predict the life duration of the vortex itself. Hence, two are the typical occurrences (figure 4.22): a positive w region is approaching to a negative one, the two w regions, opposite in sign, are departing one from each other.

In the first case (figure 4.22a), if a positive w region approaches to a negative w region, the whole vortical structure is distorted but it is not stretched. Hence, the vortical structure has more chances to survive during the motion from the impingement location. On the other hand, in case of occurrence of stretching effect, this makes the vortical structure to experience an adverse condition to travel much along the wall.

The superimposition of a tangential motion on an impinging round jet involves new interesting phenomena. Figure 4.23 reports two snapshots of the three-dimensional flow field generated by an impinging low swirling jet.

The main expected feature characterising an impinging swirling jet is the occurrence of noticeable effects related to the azimuthal component of velocity. As a matter, this is confirmed by the slight streamlines' slope with respect to the x direction in proximity of the impingement centre (red streamlines in figure 4.23a). The streamlines approaching to the wall smoothly curve and then, once reached the wall, proceed radially. A similar behaviour is noticeable for the streamlines located at the outer position (in grey colour). Not surprisingly, the slope of such streamlines, caused by rotating motion, is weaker at farther location from the jet axis. This can be explained as the superimposition of two phenomena: the spreading of the jet and the viscous shear layer effect. The first one slightly reduces the axial velocity because of the increasing of the cross-sectional area. The second phenomenon significantly reduces the azimuthal component of velocity moving away from the impingement centre.

A non-zero azimuthal velocity component does not affect only the flow field in proximity of the impingement centre. As a matter, the rotational motion significantly affects also the wall jet development. The three-dimensional black marked streamline shows the simultaneous effect of the annular vortical structures and the azimuthal velocity component. The streamline departs from the upper part of the investigation volume $(x = x_{HB}^{W})$ and rapidly deviates toward the wall. Once reached the wall, the streamline path is directed away from the impingement centre. At $0.5D$ from the origin, the streamline shows the effect of the vorticity of an annular coherent vortex. The vortical field induced by such a vortex makes the streamline to turn three times around the vortex extension before leaving it. In addition, the streamline path shows the presence of a significant negative w component which causes the azimuthal deviation of the streamline from the radial direction.

As discussed before, the relative positioning of positive and negative azimuthal velocity regions is of fundamental importance to make predictions about the life of an annular vortex. Hence, it is particularly interesting to focus the attention about the contour distribution of w over the isosurfaces. The side of annular vortex with radius equal to $1.5D$ located at $y/D \in [0.5, 1.5]$ is characterised by negative w values as shown

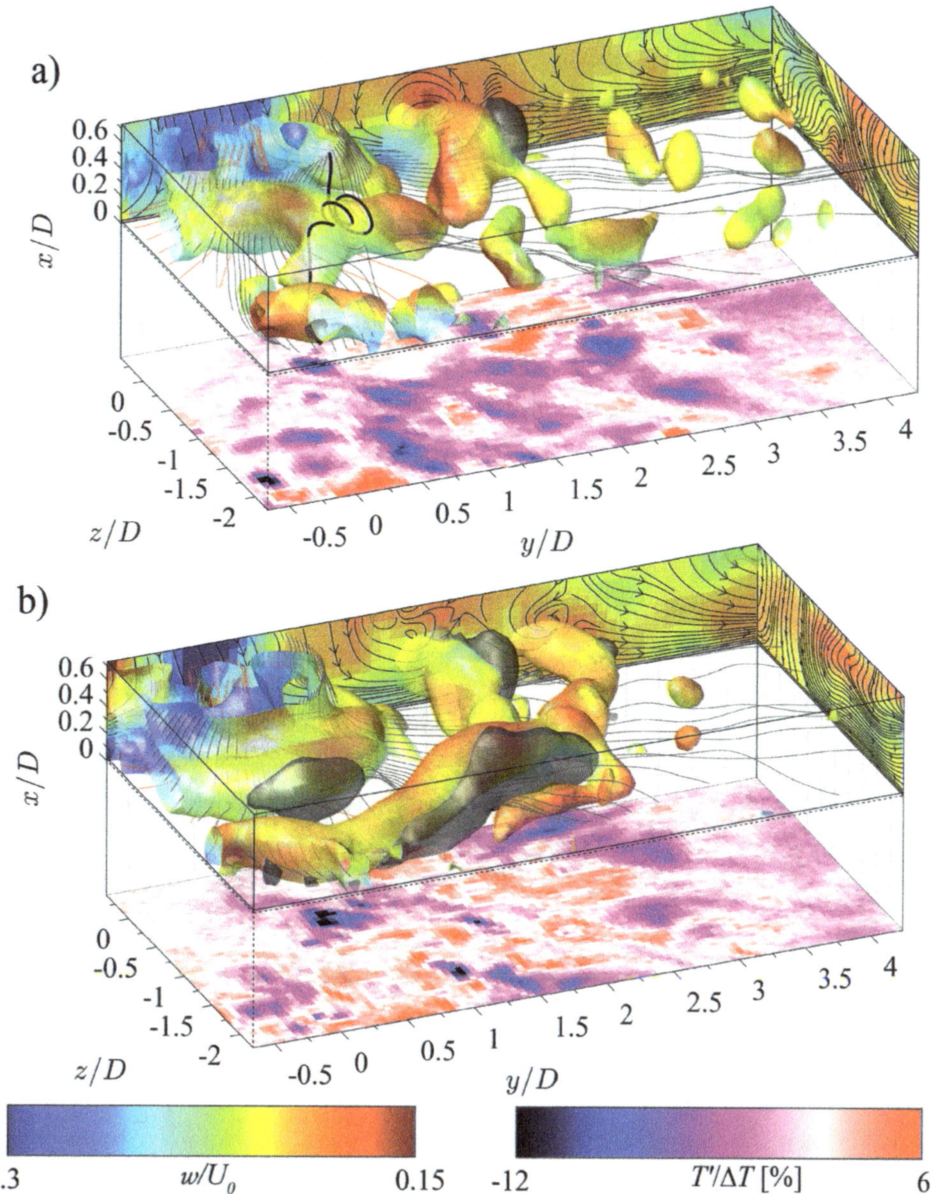

Figure 4.23: Three-dimensional representation of instantaneous flow fields of the swirling jet ($S = 0.23$) suspended at $H/D = 2$ over a flat wall. The contoured isosurface corresponds to $Q = 0.3$. The left colorbar is related to azimuthal component of velocity. The grey isosurfaces are related to $u/U_0 = 0.1$. The second colorbar is related to temperature fluctuation. The red and grey streamlines start from two concentric circumferences located at $x = x_{HB}^W$ centred on the jet axis and of radii respectively equal to $0.2D$ and $0.75D$.

by the thick black marked streamline (figure 4.23a). On the other side, the annular vortex is experiencing a positive w velocity (red contoured isosurface located at $y/D \in [1.5, 2.5]$). As reported in figure 4.22, this is the short life condition for a vortex. Indeed, it is noticeable the incoming separation in the vortical structure.

Because of the break up of the structures in the current snapshot, it is hard to find large positive u regions (grey isosurfaces). On the other hand, it is worth noting that, also at large distance from the impingement, the black streamlines located at $z = z_{HB}^{W}$ show that the fluid is dragged towards the wall because of the entrainment phenomenon. This suggests that the flow field is dominated by the entrainment phenomenon which acts the main role with respect to the fluid suction from the wall provided by the small positive u regions.

The temperature fluctuation map located below in figure 4.23a provides additional information about the temperature contribution given by each vortical structure. In proximity of the impingement centre the positive fluctuation values suggest that, with respect to the time-averaged value, the jet mainly feeds this region bringing new hot fluid. Moving away from the origin, it is worth noting the presence of two spots at opposite fluctuating values. The region located at $\{y/D \times z/D\} \in \{[1.5, 2.5] \times [0, 0.5]\}$ can be ascribed to the presence of the annular vortex and to the related positive u region. The black streamlines located on the wall at $z = z_{HB}^{W}$ show the counter-clockwise rotation of the main vortex and the drag effect involved by the u positive region. The other vortical structures which are not strong enough to drag fluid from the wall are characterised by negative temperature fluctuation values thus suggesting that the entrainment acts the main role in these cases.

Figure 4.23b shows a different self-organisation of the fluid flow. The snapshot reports three concentric annular vortices located at $1D$ of distance between one to each other. All of them are characterised by the presence of large positive u regions related to the drag effect caused by the annular vortices.

A particularly interesting phenomenon is the interaction between the two annular vortices located away from the impingement centre. A possible explanation is that the bigger vortex, which is located farther from the origin, is moving slower because of its large dimensions. Hence, the vortex closer to the origin, which is smaller, moves faster, and rapidly reaches the bigger one. The resulting interaction between the vortices provides a single structure which moves even slower but which is more unstable. Additional information can be deduced from the black streamlines in $y/D \in [1,3]$ located at $z = z_{HB}^{W}$ on the contoured boundary of the investigation volume. Two interacting rolling vortical structures can be identified at $2D$ and $2.5D$ from the origin. The upward u contribution provided by the first ring-shaped vortex directly feeds the following one. The second annular vortex pushes downward the fluid at upstream location. At downstream location, the same vortex pushes upward the fluid which is passed below the vortex. In addition to such a complex dynamics, there exists a thin layer of fluid that runs below the two vortices and which is never dragged upward from the wall.

Increasing the swirl ratio from low ($S = 0.23$) to moderate ($S = 0.43$) values involves different phenomena which were too weak to be observed in the previous cases. In figure 4.24a it is noticeable a spiral vortical structure which is cut by the $x = x_{HB}^{W}$ plane. The simultaneous presence of such a kind of vortex and a ring-shaped one, both located in proximity of the impingement centre ($r/D < 0.5$), causes their interaction. The spiral vortex rapidly deviates from the jet axis immediately adapting to the wall direction. Such a vortical structure can be ascribed to the precessing vortex core related to the swirling jet. Within $0.5D$ from the impact, the spiral vortex hits the annular vortex which is about to detach. This results in the gathering of the two structures. Another bigger and ring-shaped vortical structure is located at $r/D = 2.5$. Its size can be identified looking at the black streamlines on $z = z_{HB}^{W}$ wall. This structure is coupled with a large region characterised by positive u which is strong enough to generate a smaller vortical structure located immediately downstream $(z/D < -1.5)$.

Particularly interesting is the behaviour of the coloured streamlines in the volume included in $y/D \in [-0.5, 0.5]$. The grey ones pass below the large annular vortex and suddenly roll up because of the positive u region. On the other hand, the red ones, which belong to the internal circumference, are not affected by the uprising flow but they proceed undisturbed along the wall. This phenomenon can be explained referring to the location of the starting point of the streamlines. In this case, the red and the grey streamlines depart from two circumferences having radii equal to $0.5D$ and $1D$ respectively. As it can be noticed, they are larger than in the previous discussed cases to take into account the spreading of the jet with the increase of the swirl number. As a matter, the red ones, which are related to the vortex core, belong to the part of the flow which is continuously pushed down to the wall during its path away from the impact. In the reported snapshot the entrainment phenomenon is much more important than in the previous snapshots discussed. This can be deduced from the temperature fluctuations spots which are mainly negative all over the field of view.

Figure 4.24b shows a snapshot in which three relevant phenomena are involved: a recirculation region located in correspondence of the jet axis, a double spiral vortical structure and the gathering of two annular structures. The first phenomenon is noticeable thanks to the grey isosurface located in proximity of the origin. As confirmation of the spatial extension of this region, some of the more internal streamlines are noticeably distorted and curve toward the jet core instead of radially spread. This is the first case in which a recirculation bubble reaches the wall. The presence of this recirculation region could be caused by the simultaneous effect of the impact and of the swirling motion. This region brings high mixing contribution as it can be noticed by the random distribution of the high absolute values of temperature fluctuations.

The double spiral vortical structure reported is cut by $x = x_{HB}^{W}$ plane. Such a structure can be ascribed to the precessing vortex core which

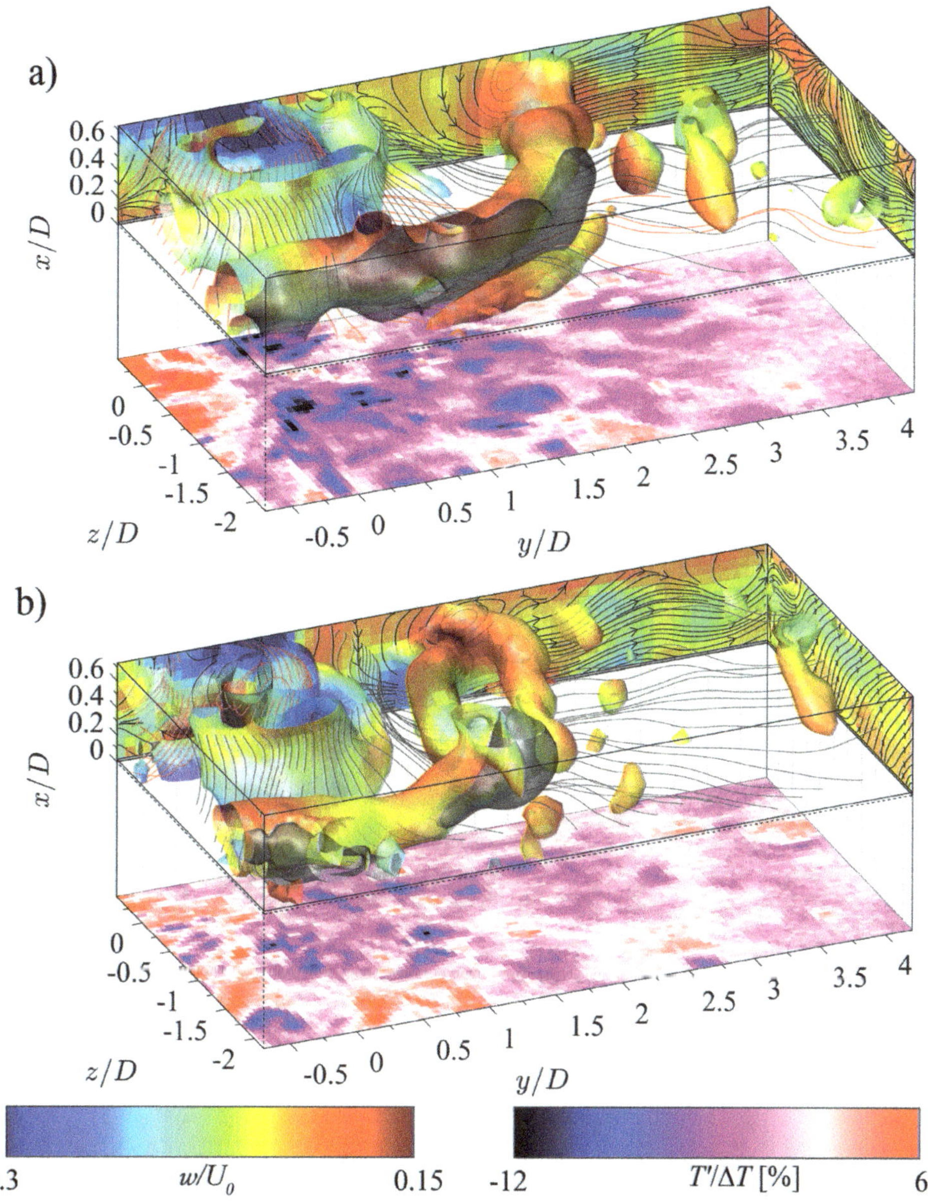

Figure 4.24: Three-dimensional representation of instantaneous flow fields of the swirling jet ($S = 0.43$) suspended at $H/D = 2$ over a flat wall. The contoured isosurface correspond to $Q = 0.3$. The left colorbar is related to azimuthal component of velocity. The grey isosurfaces are related to $u/U_0 = 0.1$. The second colorbar is related to temperature fluctuation. The red and grey streamlines start from two concentric circumferences located at $x = x_{HB}^W$ centred on the jet axis and of radii respectively equal to $0.5D$ and D.

interacts with the wall. The double spiral is smoothly deflected by the presence of the wall and the spiral begins to widen increasing its radial extension over $0.5D$.

The last phenomenon noticeable in the snapshot is the interaction between two annular vortices. The larger one, which is near to the impact, is characterised by two sections of different thickness. The first one is coupled with multiple positive u regions, the second one interacts with another half ring-shaped vortex located in the nearby. The interesting result is that the two thin vortices generate such a strong interaction that they appear as a single vortical structure as it can be noticed by the black streamlines on $z = z_{HB}^{W}$ wall.

A further increasing of the swirl ratio ($S = 0.61$) allows to better identify some of the typical vortical structures involved in a swirling jet. Figure 4.25a reports three main phenomena: two spiral vortices, a recirculation region and the break up of other two vortical structures. The double spiral vortex, which can be ascribed to the precessing vortex core, is located in proximity of the impingement. Such spiral vortices surround a recirculation region.

The two annular vortices, located at $1.5D$ and $3D$ from the origin, show a partial breakdown probably caused by the presence of gradients in the azimuthal velocity component. Because of the poor coherence of such vortical structures, the positive u regions are not as large as in the previous discussed cases.

The black streamlines located on $z = z_{HB}^{W}$ plane show a large vortex which extends from $y/D = 0.5$ to $y/D = 2$. Such a vortex provides a strong contribution to the entrainment phenomenon.

As a result of the combination of the spiral vortices and of the annular vortices, the temperature fluctuations distribution is split in two regions: one half, located at positive y, is characterised by positive temperature fluctuations, the other, located at negative y, is characterised by negative ones. The first is clearly related to the jet contribution while the second is due to the entrainment contribution which feeds the flow with cold fluid.

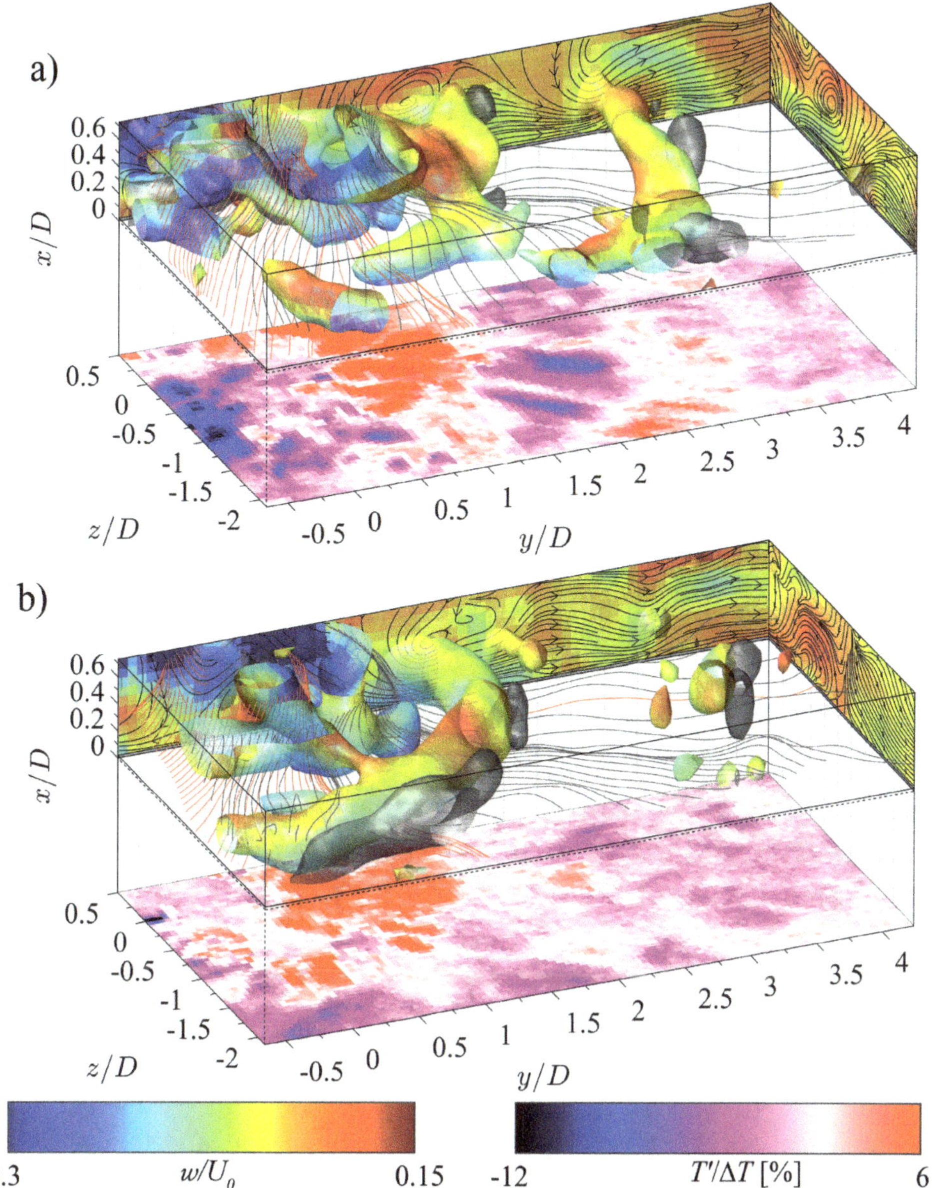

Figure 4.25: Three-dimensional representation of instantaneous flow fields of the swirling jet ($S = 0.61$) suspended at $H/D = 2$ over a flat wall. The contoured isosurface corresponds to $Q = 0.5$. The left colorbar is related to azimuthal component of velocity. The grey isosurfaces are related to $u/U_0 = 0.1$. The second colorbar is related to temperature fluctuation. The red and grey streamlines start from two concentric circumferences located at $x = x_{HB}^W$ centred on the jet axis and of radii respectively equal to $0.5D$ and D.

The coloured streamlines show the behaviour of the flow field across the investigation volume. The red streamlines, which depart from a circumference of radius $0.5D$ located on $x = x_{HB}^{W}$, show the significant rotating motion present in proximity of the vortex core. On the other hand, the grey streamlines, which depart from a circumference of radius D located on $x = x_{HB}^{W}$, are less affected by the rotating motion which rapidly decays after $1.5D$.

The snapshot reported in figure 4.25b shows how the flow field generated by a swirl number equal to $S = 0.61$ can provide enhanced mixing performances at big distance from the impingement centre. This can be explained referring first, to the flow field organisation and then, to the temperature fluctuation map.

The topology of the flow field is not much different from before. A spiral vortex and a recirculation region interacts in proximity of the impingement centre. In this case, the red streamlines are not heavily distorted. On the other hand, the grey streamlines show an heavy distortion caused by the annular vortex. They are significantly deviated from the wall direction and they roll up in correspondence of the positive u region in the volume included in $y/D \in [-0.5, 0.5]$. It is worth noting that such streamlines proceed for long distance away from the impingement. Such a behaviour explains the reason why the temperature fluctuations map is characterised by low values at large distance from the impingement. Hence, it is the result of the combined effect of both, the mixing acted by the spiral vortex in proximity of the impingement, and the cold fluid feeding role provided by the entrainment.

The last swirl ratio investigated corresponds to $S = 0.74$. The snapshot reported in figure 4.26a shows the highly turbulent character of such a flow field. The double spiral vortex is, in this case, completely dissolved after the impact, thus it is not easily recognisable as in the previous cases. As a consequence, in proximity of the impingement centre, there are small vortical structures which cannot be associated at specific patterns.

In the snapshot, two annular vortices are reported. They are located

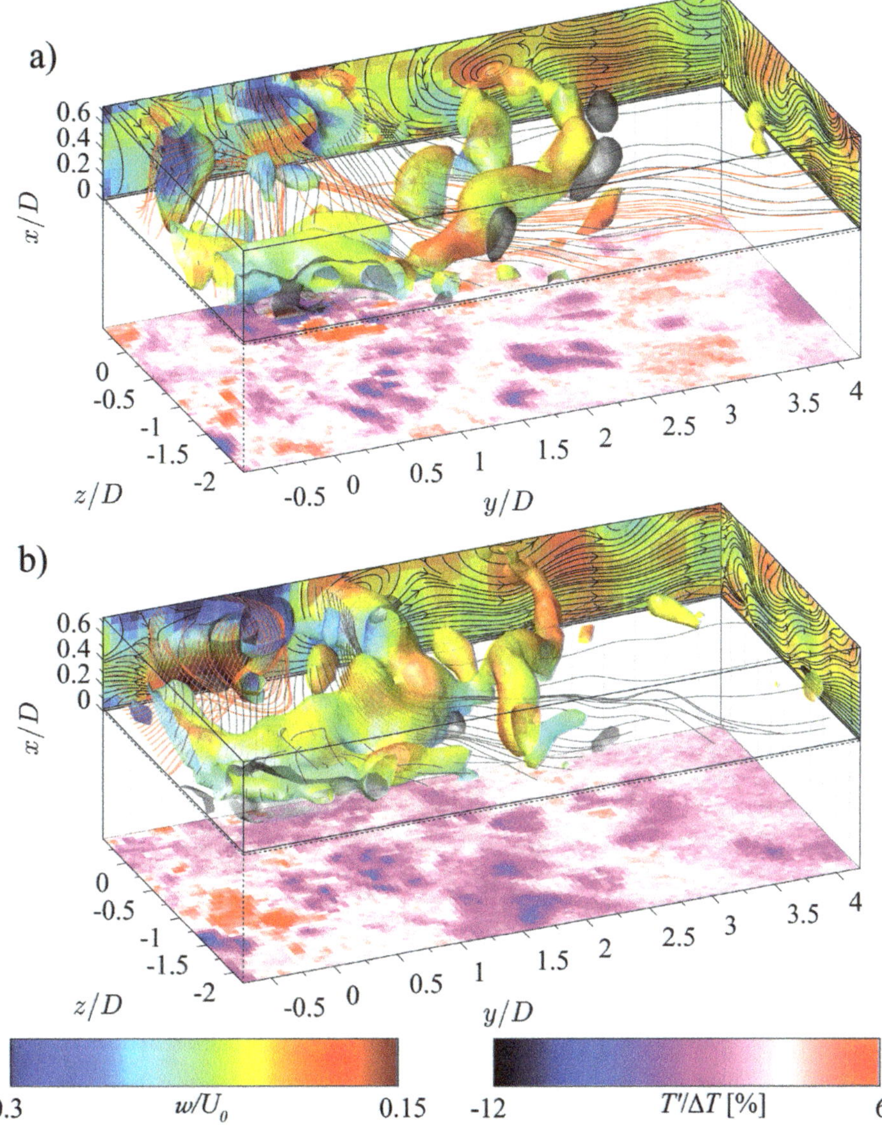

Figure 4.26: Three-dimensional representation of instantaneous flow fields of the swirling jet ($S = 0.74$) suspended at $H/D = 2$ over a flat wall. The contoured isosurface corresponds to $Q = 0.3$. The left colorbar is related to azimuthal component of velocity. The grey isosurfaces are related to $u/U_0 = 0.1$. The second colorbar is related to temperature fluctuation. The red and grey streamlines start from two concentric circumferences located at $x = x_{HB}^W$ centred on the jet axis and of radii respectively equal to $0.5D$ and D.

at a distance of $0.5D$ one from the other. The vortex located closer to the origin experiences heavy distortions because of gradients in the azimuthal velocity component. The farther one is coupled with some small positive u regions located downstream to the vortex. As discussed before, the annular vortical structures, which travel close together, appear like a single vortical structure to the main flow. This is confirmed by the black streamlines on $z = z_{HB}^{W}$ wall which show a large vortical structure which extends from $y/D = 1.5$ to $y/D = 3$. In this case, both the red and the grey streamlines proceed along the wall showing slight distortions from a radial straight path. The temperature fluctuations map faithfully imitates the annular organisation of the flow field: negative and positive ring-shaped temperature fluctuation regions alternate at $r/D = 1.5, 2, 2.5$.

The snapshot reported in figure 4.26b shows the occurrence of a large recirculation region located in correspondence of the jet axis. In this case, the red streamlines are significantly distorted. Particularly interesting is the huge vortical cluster located at $1D$ which extends to $2D$. It appears to be the result of the interaction of two or more annular vortices. The structure presents several swellings which extend outward of the main body. Each of them is characterised by a strong w gradient. At this stage two scenarios are equally possible: all of them are going to suddenly breakdown or they are just unsteady small instabilities characteristic of the chaotic motion included in the cluster. The temperature fluctuations map shows, also in this case, two ring-shaped regions respectively located at $r/D = 1.5$ and $r/D = 2$. The first one is characterised by negative fluctuation values and it is located in correspondence of the vortical cluster. The second one, mainly characterised by weak positive fluctuation values, is located in correspondence of the small u region included between the cluster and the annular-shaped vortex located at $2D$ from origin.

In the following, the POD technique is applied to both quantities acquired in the impinging configuration for two swirl numbers ($S = 0$ and $S = 0.61$) in order to extract the main vortical features involved in the velocity flow field and the patterns of the thermal field those contain the largest amount of energy.

4.2.7   POD analysis of the velocity flow field

Figure 4.27 reports the energetic contribution of the POD modes related to the velocity field of an impinging swirling jet for two different swirl number: $S = 0$ and $S = 0.61$. In the inset the cumulative sum of the energetic contributions is reported. The first four modes are selected and discussed.

In figure 4.28 the Q isosurfaces of the first four POD modes are reported. They are represented two by two in descendent order of energy content. The most energetic mode of the couple is contoured in red, while the less energetic one is coloured in blue. As it can be noticed, the first two modes (figure 4.28a) are characterised by three main vortical structures. They are all located in the region far from the impingement

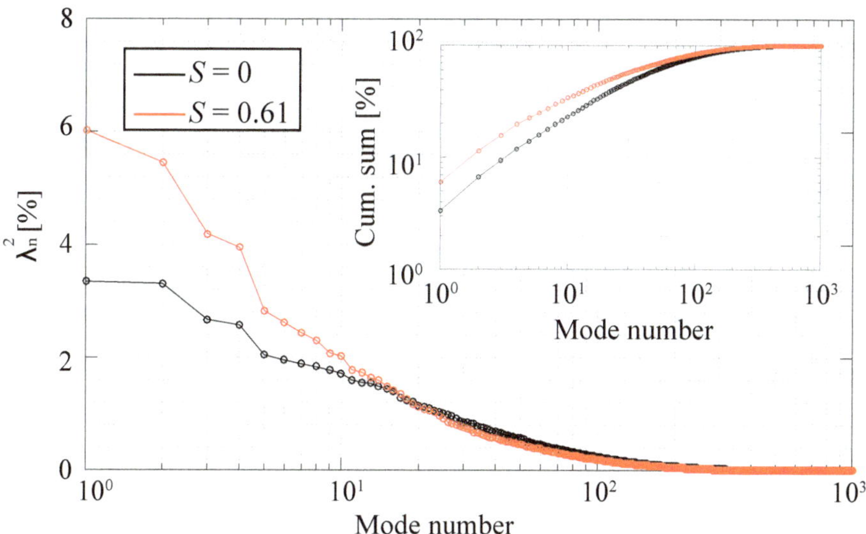

Figure 4.27: Energetic contribution of the POD modes versus the mode number in the velocity fluctuations basis for the no swirling case (black line) and the swirling case corresponding to $S = 0.61$ (red line). The inset shows the cumulative sum of the energetic contributions.

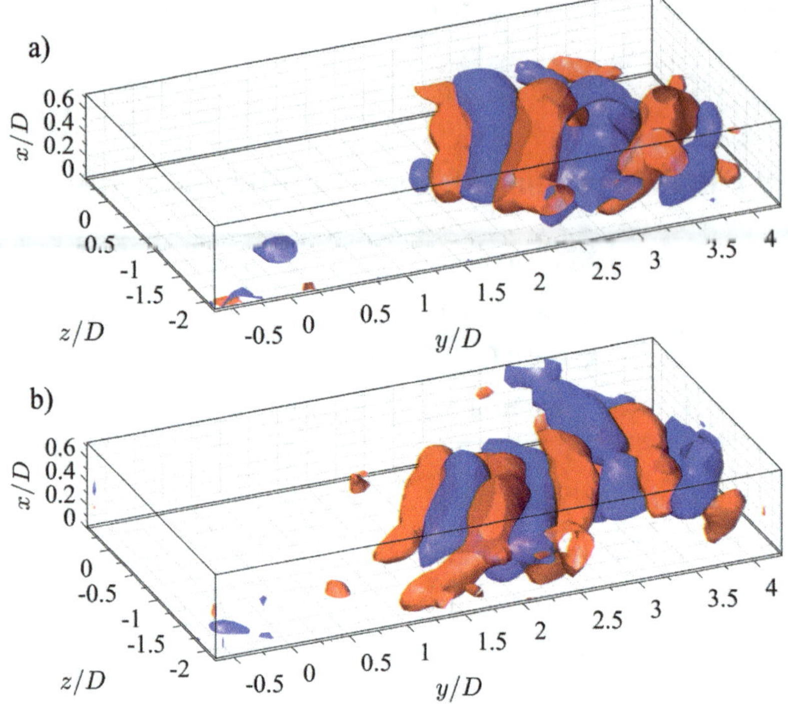

Figure 4.28: Q isosurfaces of the first four POD modes of the velocity field of an impinging circular jet: a) 1ˢᵗ and 2ⁿᵈ POD modes (respectively in red and blue colour): 6.65% of total energy; b) 3ʳᵈ and 4ᵗʰ POD modes (respectively in red and blue colour): 5.24% of total energy.

($r/D \in [1.5, 4]$) and they contain about 6.5% of total energy included in the sampled snapshots. The isosurfaces are spaced each $0.5D$ one from each other and they show a reduction in size at regions farther than $2.5D$ from the origin. It is worth noting that all the isosurfaces reported in this section were obtained with the same threshold value.

The third and the fourth modes (figure 4.28b) are characterised by a ring-shaped pattern organisation. In this case, the spacing between two consecutive structures belonging to a single mode is smaller with respect to the spacing characteristic of the first two modes. In addition, the structures are significantly smaller in thickness. This suggests that the third and the fourth modes are related to vortical disturbances which are characteristic of a higher radial spatial frequency.

In conclusion, the first four modes presented are representative of the

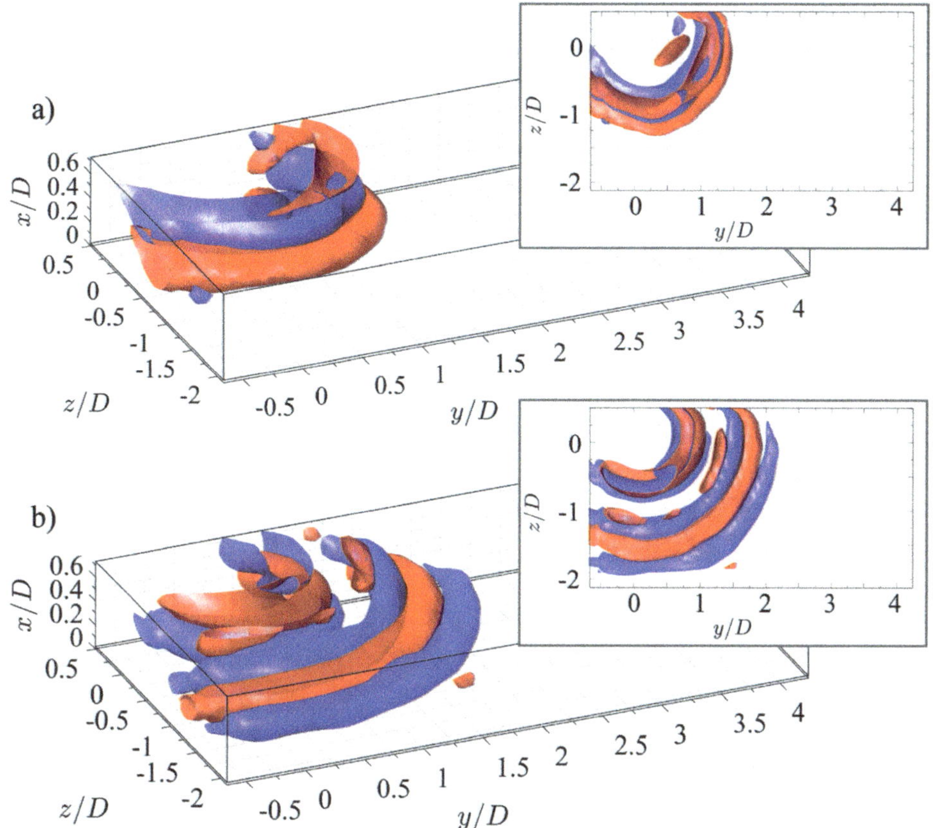

Figure 4.29: Q isosurfaces of the first four POD modes of the velocity field of an impinging swirling jet ($S = 0.61$): a) 1st and 2nd POD modes (respectively in red and blue colour): 11.46% of total energy; b) 3rd and 4th POD modes (respectively in red and blue colour): 8.14% of total energy. In each inset the top view of the related sub-figure is reported.

flow field development along the wall once the flow has deviated from the vertical direction.

The introduction of a non-zero swirl number ($S = 0.61$) significantly affects the flow field of an impinging jet. Due to the additional complexity, two views will be presented of the first four POD modes (figure 4.29). Each inset reports the top view of the related sub-figure. All vortical structures identified have a ring-shaped pattern. In addition, they are all located in proximity of the impingement region.

The first two modes reported in figure 4.29a are representative of about

11% of the total energy included in the snapshots sampled. The two rings have the same radius which is equal to D. They can be explained as the consequence of the interaction of the precessing vortex core with the wall.

The second couple of modes shows a different behaviour (figure 4.29b). A thinner annular-shaped pattern is located at $0.5D$ while another vortical pattern is located at $r/D \in [1, 2]$. Hence, the structures included within the nozzle extension can be ascribed to the inner secondary vortex impacting on the wall, the larger ones are related to the outer secondary vortex which significantly increases its radius to smoothly interact with the wall.

4.2.8   POD analysis of the thermal field

The POD analysis of the thermal footprint of an impinging swirling hot jet exhausting in a cold ambient is discussed in the following. Two different swirl numbers are investigated: $S = 0$ and $S = 0.61$.

In the present experimental apparatus, the assumption according to which the temperature in proximity of the jet impingement is constant during the whole duration of the thermal acquisition (about 75s) is not correct. For this reason, it is fair assuming the presence of a slight temperature drift. Due to the complex dynamics involved, it cannot be corrected with standard detrending methods. However, such a temperature drift significantly affects the POD modes and their energy ranking. Hence, for both the swirl numbers presented, the POD modes, related to temporal coefficients whose linear regression is characterised by a non-zero first derivative, will be neglected in the following discussion.

Given the round shape of the nozzle, it is fair assuming an azimuthally independent thermal field. Hence, the field of view is cropped to a square, centred on the impingement centre. In order to better identify the main thermal features, a circular blanking mask, centred on the impingement centre, is applied to the data set. Then, in order to increase the number of snapshots employed in the POD analysis, the blanked data set is rotated three times around the impingement centre, identified with sub-pixel ac-

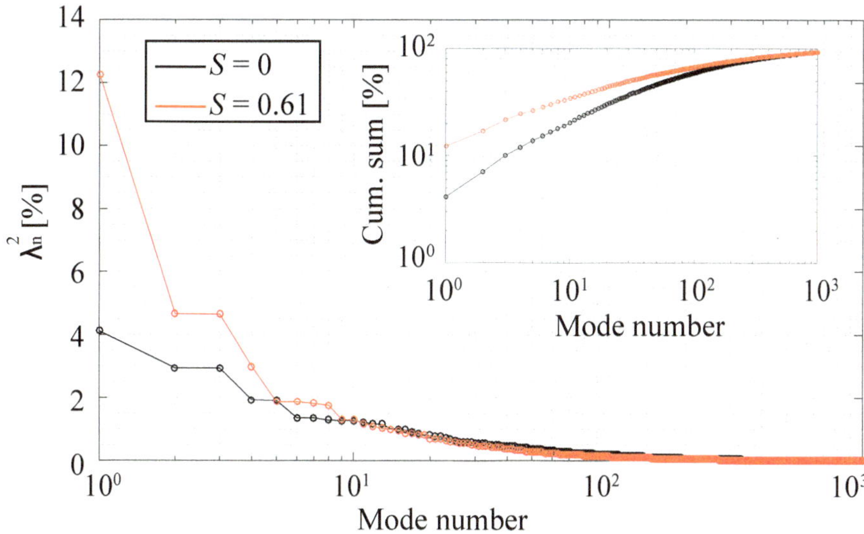

Figure 4.30: Energetic contribution of the POD modes versus the mode number in the thermal fluctuations basis for the no swirling case (black line) and the swirling case corresponding to $S = 0.61$ (red line). The inset shows the cumulative sum of the energetic contributions.

curacy. Each rotation is equal to a quarter of a round angle. These four sets of matrices are stacked in a single 3D matrix and this is assumed as the new matrix of the snapshots. As a consequence, the number of snapshots used to operate the POD analysis is multiplied by four.

Figure 4.30 shows the energy contribution of the POD modes, which is significantly different for the two swirl numbers investigated.

Figure 4.31 reports the first six POD modes of the thermal footprint of an impinging circular hot jet exhausting in a cold ambient. The modes are respectively representative of 2.94%, 2.94%, 1.93%, 1.90%, 1.35%, 1.35% of the total energy included in the snapshots analysed.

As it can be noticed, all patterns are characterised by high spatial coherence. The first two modes (figure 4.31a and figure 4.31b), which contain about 6% of the total energy, are characterised by the presence of two large patterns having spatial period equal to half of the perimeter of the circumference investigated. All the next coupled modes (figures 4.31c-f) are representative of the upper harmonic components of the first

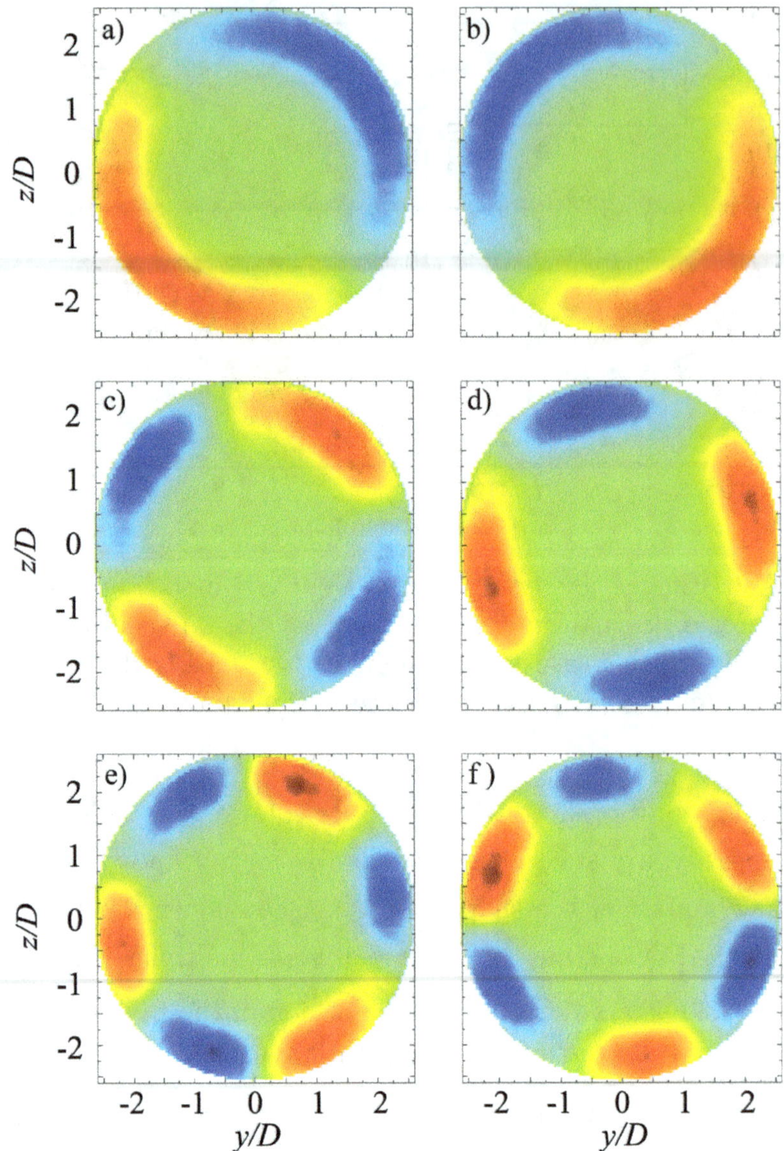

Figure 4.31: First six POD modes of the thermal field of an impinging circular jet. The modes related to the temperature drift are neglected. The modes contain respectively 2.94%, 2.94%, 1.93%, 1.90%, 1.35%, 1.35% of the total energy included in the snapshots analysed.

two modes. It is worth noting that, all features reported in figure 4.31 are located at the same distance from the impingement, i.e. $r/D = 2.25$, which corresponds to the location of the secondary vortex in an impinging circular jet. Hence, these structures are representative of the highest

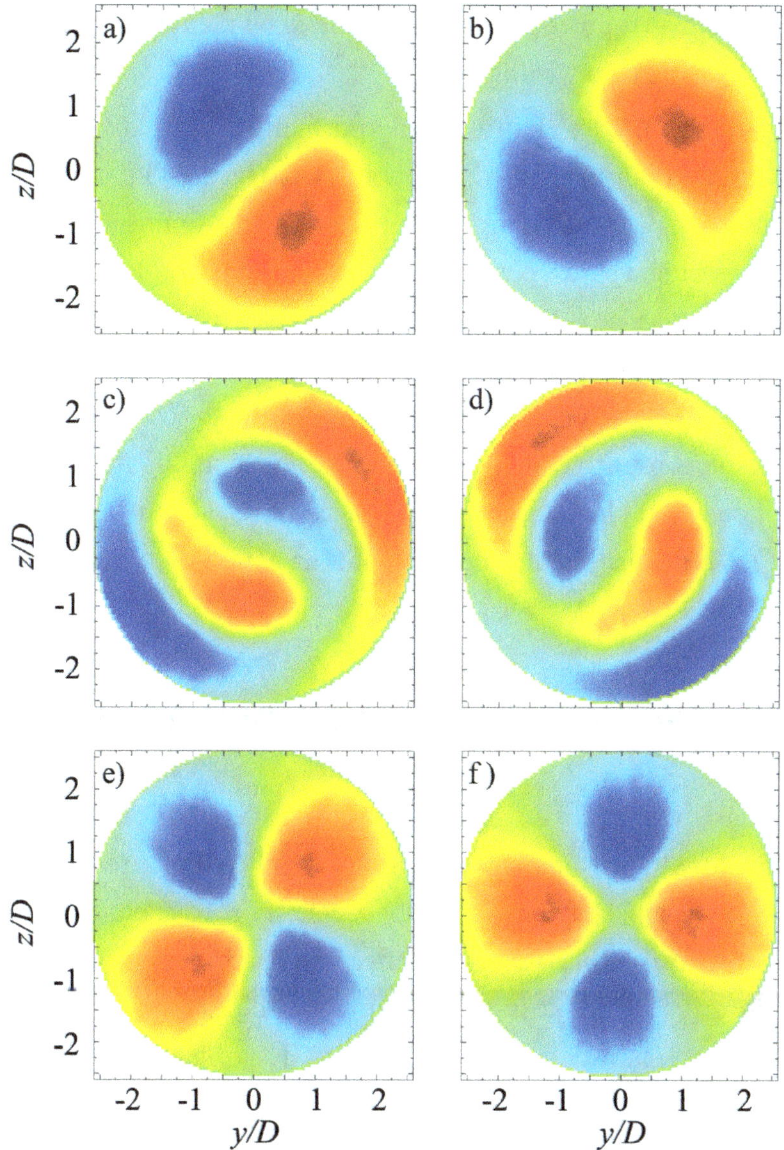

Figure 4.32: First six POD modes of the thermal field of an impinging swirling jet ($S = 0.61$). The modes related to the temperature drift are neglected. The modes contain respectively 4.65%, 4.65%, 1.86%, 1.86%, 1.81%, 1.74% of the total energy included in the snapshots analysed.

energetic contributions of the secondary vortical structures generated on the wall. For this reason, the secondary vortex can be assumed as the main actor of the underlying non-linear thermal dynamics.

In figure 4.32 the first six most energetic POD modes corresponding to $S = 0.61$ are reported. The patterns shown in this case are much different from the no swirling case.

The first two modes (figure 4.32a and figure 4.32b) are representative of the rigid rotation of the whole thermal pattern. The second coupled modes (figure 4.32c and figure 4.32d) suggest the simultaneous influence of the precessing vortex core and of the outer secondary vortex. The patterns associated to PVC are those located in proximity of the origin. They can be enveloped in a circle of radius equal to D. On the other hand, the patterns located at $2D$ from the origin are related to the OSV. The last two modes reported (figure 4.32e and figure 4.32f) have a dominant radial extension. This suggests that such patterns can be associated to the wall jet influence on the thermal fluctuation displacement.

4.3   Velocity and temperature correlations

The POD analyses reported above do not investigate the mutual influence of the two quantities measured. Hence, the results presented are not suitable to recognise a correlation between the two unsteady dynamics. As a matter, in the following two different correlation techniques will be applied in order to identify the features of the flow field which can be ascribed as the main actors of the thermal dynamics.

The correlations will be presented investigating different ranges of influence between the two measured quantities. As first, the mutual relationship between the quantities will be investigated through by the application of Extended POD (EPOD) technique (Borée, 2003). Such a modal decomposition technique assures that, given a common functional temporal basis, the modes extracted and the energy associated, will be representative of the main features involved and of their relative importance in the mutual influence between the two quantities.

The second analysis presented is the correlation between the thermal field and the closest to the wall slice of the velocity volume.

4.3.1 Extended Proper Orthogonal Decomposition

The Extended POD technique (Borée, 2003; Maurel et al., 2001) takes advantage of information provided by the two quantities simultaneously measured. As a matter, one of the key features of the present measurements is their synchronisation in time and the knowledge of relative positioning between the thermal and the velocity frames. Hence, the purpose of the application of such a technique is to find some significant global correlation between the thermal and the velocity field.

The fundamental assumption of such a procedure is that the two measurements share the same temporal correlation coefficients matrix. Hence, at this stage no assumption is needed about the relative positioning of the two frames.

Figure 4.33 reports the energetic contribution of the first thousand EPOD modes related to $S = 0$ (black line) and $S = 0.61$ (red line). The inset image is representative of the cumulative sum of such energetic contributions. As it can be noticed, the two curves do not monotonically decrease. This is a direct consequence of the application of such a decomposition method. In contrast with the energy-ranked modes organisation of POD technique, in this case the energetic contribution of each mode provides an estimation of the correlation degree existing between the thermal and the velocity fields.

As it can be noticed, the maximum energetic contribution percentages reported does not overcome 2%. This can be explained assuming that the temporal correlation matrix related to the two measured quantities is poor. Hence, it is fair thinking that a local correlation exists. For this reason, taking advantage of the relative positioning of the two frames, local correlation techniques could be applied to get deeper in the relationship between the two dynamics.

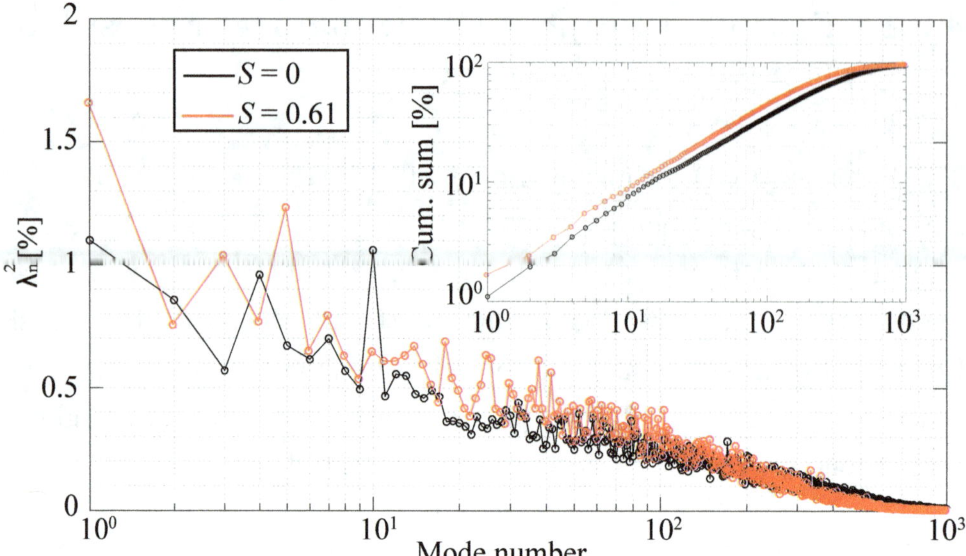

Figure 4.33: Energetic contribution of EPOD modes in the velocity basis versus the mode number for the no swirling case (black line) and the swirling case corresponding to $S = 0.61$ (red line). The inset shows the cumulative sum of the energetic contributions.

4.3.2 Partially global correlations

The normalised correlation coefficient is a dimensionless statistical quantity that estimates the strength of the relationship between the relative movements of the two variables ($\phi \triangleq \Phi + \phi'$ and $\psi \triangleq \Psi + \psi'$) from their average along a specified dimension. As previously defined, the capital letters indicate the time-averaged quantities, the lower-case letters with apex indicate the fluctuating quantities.

In the present case, the two variables are the temperature fluctuations map and one of the three velocity fluctuations extracted from the lowest slice of the three-dimensional distribution. Hence, the normalised correlation coefficient is defined as follows:

$$C_\rho(x_{LB}^W, y, z) = \sum_{t \in \mathrm{T_{sync}}} \frac{(\phi'(x,y,z,t), \psi'(y,z,t))}{\sigma_\phi(x,y,z)\sigma_\psi(y,z)N_{\mathrm{obs}}}, \quad x = x_{LB}^W, \forall(y,z) \in \Omega$$

$$(4.4)$$

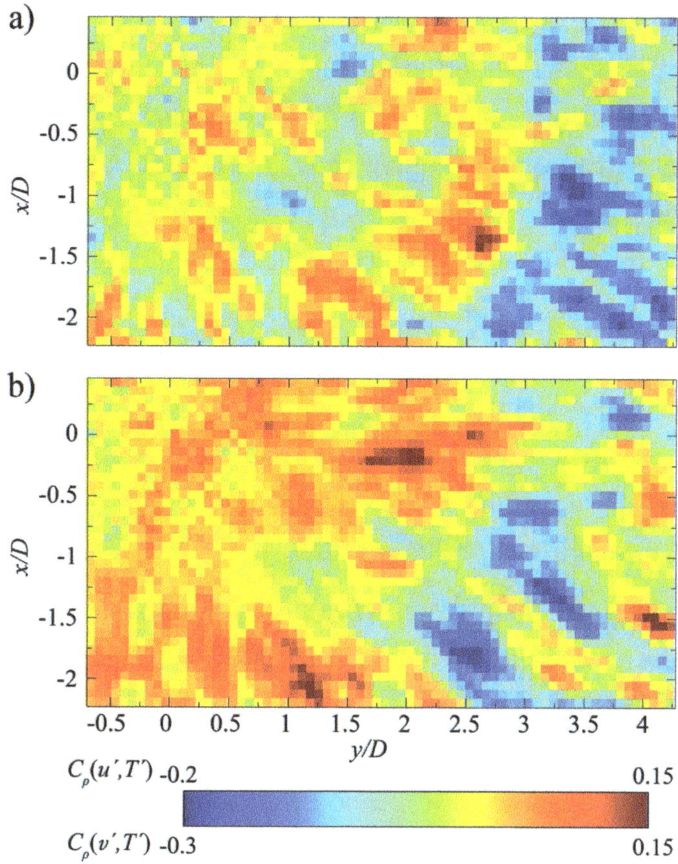

Figure 4.34: Normalised correlation maps between the slice of velocity volume closest to the wall and the temperature map for an impinging circular jet. a) $C_\rho\left(u'(x_{LB}^W, y, z), T'(y, z)\right)$, b) $C_\rho\left(v'(x_{LB}^W, y, z), T'(y, z)\right)$.

where $(\cdot, \cdot)$ indicates the dot product along the time dimension, $\sigma_{(\cdot)}$ refers to the standard deviation of the variable in brackets, Ω is the projection along the x direction of the field of view shared by thermal and velocity measurements, T_{sync} is the time duration of the synchronised test, N_{obs} is the number of observations employed. Such a quantity has a range of values bounded by 1 on an absolute basis.

Figure 4.34 reports the correlation maps related to the no-swirl case. The first frame (figure 4.34a) shows the correlation map between the fluctuations of the axial velocity component and the fluctuations of the temperature. The distribution presented identifies a high and positive correlation annular region at radial distance $r/D \in [2,3]$. This region coincides with the high fluctuation region for an impinging circular jet. A positive correlation value can be obtained when the fluctuations are positive or negative for both quantities. In the first case, a recirculation region related to the flow field corresponds to fluid spots which are hotter than the average value of temperature. The second case is possible when cold fluid spots impact on the wall. This can be ascribed to the entrainment phenomenon which drags new fluid toward the direction of the jet axis. Both of these phenomena are possible and they can be both related to the mixing phenomena occurring approximately at $2.25D$ from the origin. This is in accordance with the location of maximum of standard deviation of temperature profile reported in figure 4.14.

Another annular region is noticeable in the correlation map. It is located at the boundary of the investigation domain. Such a high negative correlation region located at $4D$ from the origin suggests a dual phenomenology. In this case cold spots of fluids are in correspondence of recirculation regions or, vice versa, hot fluid spots are impacting the wall. These two phenomena are not equally possible. As a matter, the first occurrence is a consequence of the residual annular structures moving away from the impinging centre (figure 4.20). They continue raising up fluid from the wall which is issued by jet hot fluid. On the other hand, the second possibility is very difficult to occur. As a matter, this would mean that hot spots belonging to the jet are impacting again far away from the impingement centre.

The radial component of velocity shows a different overall behaviour (figure 4.34b). A large high positive correlation region can be identified in proximity of the impact. Such a region has radius equal to $2D$ and identifies the location where the radial component of velocity fluctuations is positive in correspondence with positive thermal fluctuations. The opposite condition, occurrence of cold fluid spots moving toward the impact, would be not physical due to the proximity to the impingement region.

Another characteristic zone is noticeable, it is located at radial distance $r/D = 3$ from the origin. This highly negative correlation region suggests two possible occurrences: negative thermal fluctuation regions which are unsteadily pushed away from the impact or positive thermal fluctuation regions which are unsteadily dragged back to the impact. The first of these explanations appears to be fairer because of the correspondence of such a region with the high fluctuation region located at $2.25D$ from the impingement for a no swirling jet.

The superimposition of a tangential motion to a circular jet involves different behaviours with respect to the previous case. Figure 4.35a shows the correlation $C_\rho(u', T')$ for an impinging swirling jet ($S = 0.61$). The correlation map related to the axial component of velocity fluctuation is characterised by two concentric regions. The smaller one, which has radius equal to $1.5D$, is characterised by negative correlation values. It can be ascribed to the precessing vortex core which brings hot portions of fluid towards the wall. On the other hand, the large positive correlation annular region located at radial distance $r/D = 2.25$ from the origin demonstrates that the contribution to correlation of axial component of velocity fluctuations provided by a swirling flow is comparable with the conventional case.

A significant difference from the no swirling case can be noticed in the correlation map related to the radial component of velocity fluctuations (figure 4.35b). As a matter, the superimposition of a tangential motion causes the occurrence of two regions, a negative one located in proximity of the impact and a positive one located at radial distance within the range $r/D \in [2, 3.5]$. The first region can be ascribed to a recirculation bubble which makes positive temperature fluctuation spots to unsteadily move backward to the impingement. The second region shows positive correlation values far from the impact. This can be ascribed to portions of fluid characterised by positive temperature fluctuations which travel along the wall passing below the annular vortical structures crawling away from the impact.

Figure 4.35c is representative of the correlation between the fluctua-

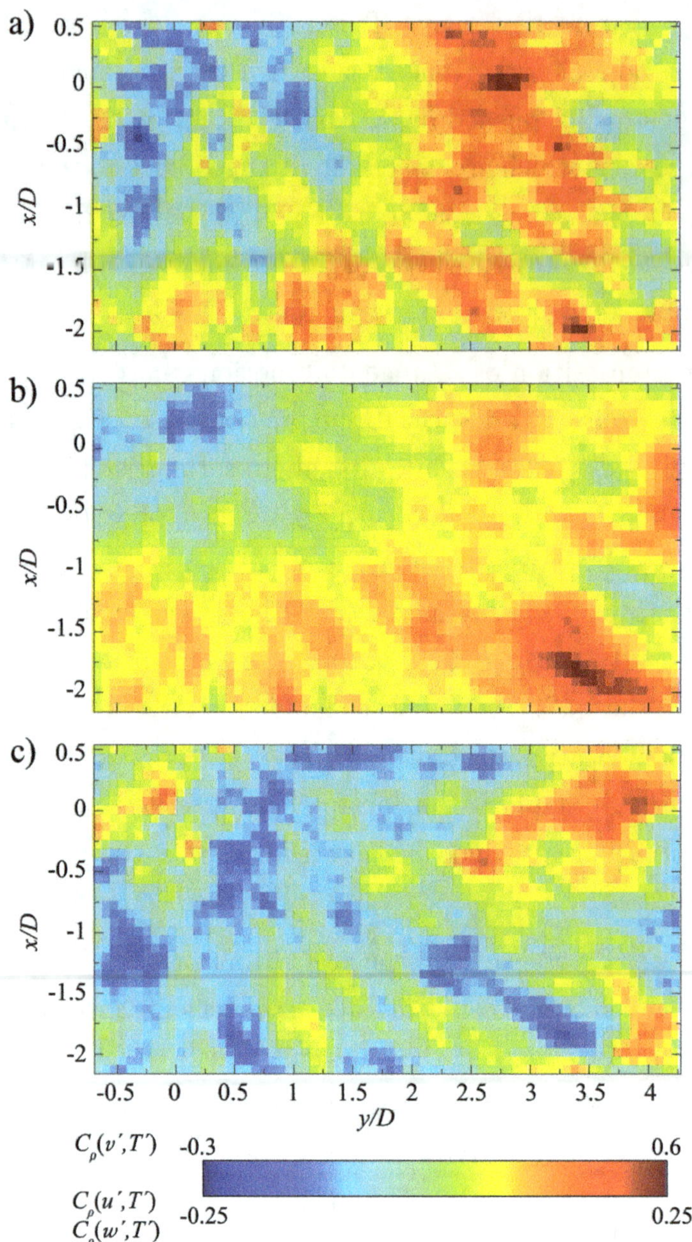

Figure 4.35: Normalised correlation maps between the slice of velocity volume closest to the wall and the temperature map for an impinging swirling jet ($S = 0.61$). a) $C_\rho\left(u'(x_{LB}^W, y, z), T'(y, z)\right)$, b) $C_\rho\left(v'(x_{LB}^W, y, z), T'(y, z)\right)$, c) $C_\rho\left(w'(x_{LB}^W, y, z), T'(y, z)\right)$.

tions of azimuthal component of velocity and the temperature fluctuations $(C_\rho(w', T'))$. It is worth remembering that the swirling sense imposed has clockwise orientation in the reference system adopted (figure 3.10). Hence, referring to the $\overline{w'^2}$ distribution reported in figure 4.13, it is possible to relate the large negative correlation region with the radial extension of the positive value assumed by the $\overline{w'^2}$. As a matter, hot portions of fluid are characterised by clockwise azimuthal fluctuations of velocity. The noticeable large positive correlation region located far from the origin can be ascribed to a residual effect of the entrainment of cold fluid at large distance from the impingement.

4

5 | Conclusions

In the present work time-resolved Tomographic PIV and Infrared thermography are simultaneously applied to investigate the effect of low to high swirl number on the mixing and the transport of passive scalars in an impinging hot jet exhausting in a cold ambient.

A custom swirl generator is designed to satisfy three different requirements: fine tuning of the swirl ratio, fast switching among different swirl numbers, absence of features in the flow which resemble the swirl generation method. The device includes three 3D-printed components, two of them build the housing of the swirl generator which is a disk with equally spaced vanes.

The experimental setup designed to test such a device is a water facility made of two tanks at different temperatures. The tank which feeds the swirl generator is filled with hot water, the other, in which the test is conducted, is provided of a visible and IR transparent optical access on the bottom which acts as impingement surface. Such an experimental arrangement is suitable to simultaneously acquire the three-dimensional flow field and the thermal footprint of the swirling impinging jet.

The time-averaged velocity profiles for nine vane angles are investigated in proximity of the nozzle. Then, the swirl number S related to each of the swirl generator designed is estimated applying the most general definition of swirl number (Toh et al., 2010). The swirl number increasing affects all the components of the velocity. The axial component of velocity experiences a gradual defect along the jet axis. The jet spreading becomes significant from moderate ($S = 0.43$) to high ($S = 0.74$) swirl numbers. The azimuthal velocity component and the circulation profiles show a strong dependence of the swirl numbers on the vane angles.

The unsteady three-dimensional behaviour of the flow field in proximity of the nozzle is reported for five of the swirl generators designed. The vortical structures completely change with the introduction of a weak swirling motion ($S = 0.23$): a spiral pattern embraces a central vortical column. Increasing the swirl number from low to moderate ($S = 0.43$) values, the vortical structure becomes thicker and the vortex core begins to be significantly affected by the spreading. The snapshots related to the last two swirl numbers presented ($S = 0.61$ and $S = 0.74$) show another typical vortical structure involved in the free flow of a swirling jet: the precessing vortex core (Cala et al., 2006). Such a structure rotates around the jet axis in-phase with the outer vortex. The high rotation ratio causes the occurrence of a recirculation region aligned with the jet axis.

The application of the POD analysis on the free flow field of a swirling jet shows the three most energetic features involved: the precessing vortex, the inner secondary vortex and the outer secondary vortex.

In the second part of the thesis, the impinging configuration of a swirling jet is investigated. The time-averaged velocity fluctuation profiles at the wall show the formation of a recirculation bubble in proximity of the impact region. The size of such a region non-linearly depends on the swirl number. The time-averaged temperature fluctuation profiles suggest that the mixing performances of a swirling jet are significantly higher than those provided by a conventional jet. In addition, the maximum location of the temperature fluctuations significantly moves depending on S.

Additional information about the mixing performances of an impinging swirling jet can be deduced from the time sequences of temperature fluctuations at several stations from the impingement centre. A conventional no swirling jet shows poor mixing performances until $1.5D$. On the other hand, the swirling motion is demonstrated to be an entrainment trigger since $0.8D$ from the impingement centre. The mixing performances driven by the swirling motion are clearly highlighted by the temperature fluctuation time sequences estimated at the radial distance of $2.3D$. At this stage, $S = 0.61$ shows the best mixing performances.

Taking advantage of the displacement in time of temperature fluctuation regions, thermal images are processed with a cross-correlation algorithm. Such a procedure provides the time-averaged profile of the radial component of velocity at $x = 0.01D$ from the wall. The comparison of such a distribution with V component extracted from the lowest height of Tomographic PIV volume for $S = 0$ case provides the approximate behaviour of velocity profiles along the radial direction. The radial velocity evaluated by PIV technique experiences a significant acceleration at $0.65D$ from the impact. Such a velocity variation is smoothly transmitted to the thermal field.

The discussion about the unsteady behaviour of an impinging jet is completed by the three-dimensional visualisation of the vortical structures involved. In correspondence of the velocity field, the simultaneously acquired fluctuation map is reported. The no swirling jet is characterised by ring-shaped vortical structures. Several recirculation regions are noticeable. Significant temperature fluctuation values are all located farther than $1.75D$ from the origin.

The introduction of a swirling motion increases the extension of the recirculation regions downstream of the ring-shaped vortices. In addition, it is noticeable the occurrence of regions which tend to rotate in opposite sense. The distance that the ring-shaped vortices can travel away from the impingement centre strongly depends on such local azimuthal component gradients.

In order to identify the main features of the two separate fields, the POD analysis is applied to the impinging configurations for two swirl numbers: $S = 0$ and $S = 0.61$. At first, the POD analysis of the velocity field is performed. The modes related to the no swirling case show to have a ring-shaped vortical structure. They are located at radial distances of $1.5D$ from the origin. On the other hand, the swirling case shows more energetic vortical structures mainly located in proximity of the impact region.

The POD analysis of the thermal data is performed on a squared frame with the application of a circular blanking mask in order to better identify the main features involved. The patterns related to the modes of a no swirling jet show to be mainly located at $2.25D$ from the origin. Such a location corresponds to the secondary vortex in a conventional impinging jet. On the other hand, the patters related to the modes of the thermal field of an impinging swirling jet can be ascribed to three main features: the rigid rotation of the whole thermal pattern, the simultaneous influence of the PVC and of the OSV, the displacement of the thermal fluctuation regions along the radial direction.

The analyses discussed above do not take advantage of the synchronisation and of the relative positioning of the frames related to the two measured quantities. The assumption of simultaneous acquisition is used to apply the Extended POD technique. The results show poor global correlation between the two data sets. This suggests that the two phenomena produce a poor temporal correlation matrix. Hence, a different local correlation logic is applied: the normalised correlation coefficient maps between the temperature fluctuations and all components of velocity fluctuations located on the lowest slice of the velocity volume are estimated. For the no swirling case a large high correlation region, related to the fluctuations of the axial velocity component, is located at $r/D \in [1.5, 2.5]$. On the other hand, a high correlation region related to the fluctuations of the radial component of velocity is located in proximity of the impingement.

The correlation maps related to the swirling case show specific regions of influence for each component of velocity fluctuation. The correlation map related to the axial component of velocity fluctuations shows two high correlation annular regions centred on the origin. The small one can be ascribed to the PVC which brings hot portions of fluid towards the wall. The large one, located at radial distance of $2.25D$ from the origin resembles the pattern provided by the conventional round jet.

The correlation map related to the radial component of velocity fluctuation shows two characteristic regions. The small negative one, located in proximity of the impact, can be ascribed to a recirculation bubble which unsteadily acts to drag hot fluid backward to the impingement. The positive large one can be ascribed to portions of fluid characterised by positive temperature fluctuations which pass below the annular vortical structures crawling away from the impact.

The last correlation map reported, related to the azimuthal velocity component fluctuations, shows a large high correlation region which, in accordance with the non-zero time-averaged radial extension of the $\overline{w'^2}$ distribution, suggests that portions of fluid characterised by positive temperature fluctuations are moved accordingly to the sense of swirl rotation.

5

Appendices

A Basic equations for the swirling jet

The theoretical description of the velocity distribution of a free swirling jet is based on the Reynolds-averaged Navier-Stokes equations written in a cylindrical inertial frame (x, r, θ). The reference system adopted is reported in figure 1.13. The velocity components in cylindrical coordinates will be denoted with (u, v, w). The following mathematical treatment, adapted from Shiri (2010), is based on two fundamental assumptions:

- the mean flow is axisymmetric;

- the equations are written at large distances from the jet source.

The first assumption assures that the description of the flow field can be operated by the investigation of the radial profiles of velocity. Mathematically, this means that the azimuthal derivative of any averaged quantity is identically equal to zero.

The second assumption assures that the time-averaging and ensemble averages are the same in the limit of an infinitely long record and an infinite number of members of the ensemble:

$$\lim_{T_{\text{obs}} \to \infty} \frac{1}{T_{\text{obs}}} \int_0^{T_{\text{obs}}} \psi \, dt = \Psi \tag{A.1}$$

where T_{obs} is the period of the observation, ψ is a generic statistical independent quantity involved in the flow field evolution and Ψ is the time-averaged quantity. In the following, a flow field velocity component distribution will be denoted with a lowercase letter, the capital letters will indicate the time-averaged quantities, the apex on lowercase letters will indicate the fluctuating quantities.

A.1 Mass conservation

In absence of mass sources, the continuity equation is the same for both the swirling and the non-swirling jet. Assuming constant density flow, the Reynolds-averaged continuity equation is given by:

$$\frac{\partial U}{\partial x} + \frac{1}{r}\frac{\partial}{\partial r}rV + \frac{1}{r}\frac{\partial W}{\partial \theta} = 0. \tag{A.2}$$

Applying the condition that the flow is axisymmetric, the last term in equation (A.2) is identically to zero, i.e. $\partial/\partial\theta = 0$. Thus the Reynolds-averaged continuity equation results in:

$$\frac{\partial U}{\partial x} + \frac{1}{r}\frac{\partial}{\partial r}rV = 0. \tag{A.3}$$

A.2 Momentum conservation

For high Reynolds numbers, the viscous terms of the momentum Reynolds-averaged-Navier-Stokes equations in cylindrical coordinates can be neglected. In addition, assuming azimuthal symmetry, it can be shown that such equations result in:

$$U\frac{\partial U}{\partial x} + V\frac{\partial V}{\partial r} = -\frac{1}{\rho}\frac{\partial P}{\partial x} - \frac{\partial \overline{u'v'}}{\partial r} - \frac{\overline{u'v'}}{r} - \left\{\frac{\partial \overline{u'^2}}{\partial x}\right\} \tag{A.4}$$

$$-\frac{W^2}{r} = -\frac{1}{\rho}\frac{\partial P}{\partial r} - \frac{\partial \overline{v'^2}}{\partial r} + \frac{\overline{w'^2} - \overline{v'^2}}{r} - \left\{\frac{\partial \overline{u'v'}}{\partial x}\right\} \tag{A.5}$$

$$U\frac{\partial W}{\partial x} + V\frac{\partial W}{\partial r} + \frac{VW}{r} = \frac{\partial \overline{u'v'}}{\partial r} - 2\frac{\overline{u'v'}}{r} - \left\{\frac{\partial \overline{u'w'}}{\partial x}\right\} \tag{A.6}$$

where P is the time-averaged static pressure. Typically, for a stationary free jet the last terms in brackets of equations (A.4) to (A.6) are neglected since they represent changes in the streamwise direction. The mixed second-order moments can be reasonably neglected because smaller than the pure ones. Equation (A.5) can be integrated from a radial value r to ∞ to explicitly write the pressure term:

$$\frac{1}{\rho}\left[P(x,r) - P(x,\infty)\right] = -\int_r^\infty \frac{W^2 + \left[\overline{w'^2} - \overline{v'^2}\right]}{\tilde{r}}d\tilde{r} - \overline{v'^2} \tag{A.7}$$

where $\tilde{r}$ is the integration variable along the radial direction. In equation (A.7), W and all turbulence quantities are supposed to vanish at infinite radius. Differentiating the equation (A.7) with respect to x, results in:

$$\frac{1}{\rho}\frac{\partial P(x,r)}{\partial x} = \rho\frac{dP(x,\infty)}{dx} - \frac{\partial}{\partial x}\int_r^\infty \frac{W^2 + \left[\overline{w'^2} - \overline{v'^2}\right]}{\tilde{r}}d\tilde{r} - \frac{\partial\overline{v'^2}}{\partial x}. \quad (A.8)$$

If it is assumed that no external flow is present, then $dP(x,\infty)/dx = 0$. Hence, substituting equation (A.8) into equation (A.4) yields the streamwise momentum equation:

$$U\frac{\partial U}{\partial x} + V\frac{\partial U}{\partial r} = -\frac{1}{r}\frac{\partial r\overline{u'v'}}{\partial r} + \frac{\partial}{\partial x}\left\{\left[\overline{v'^2} - \overline{u'^2}\right] + \int_r^\infty \frac{W^2 + \left[\overline{w'^2} - \overline{v'^2}\right]}{\tilde{r}}d\tilde{r}\right\}.$$

$$(A.9)$$

A.3   Integral equations of axial and angular momentum conservation

In order to obtain the axial momentum in axial direction, the equation (A.3) should be multiplied by U and added to equation (A.9) yielding to:

$$\frac{\partial U^2}{\partial x} + \frac{1}{r}\frac{\partial rUV}{\partial r} = -\frac{1}{r}\frac{\partial r\overline{u'v'}}{\partial r} + \frac{\partial}{\partial x}\left\{\left[\overline{v'^2} - \overline{u'^2}\right] + \int_r^\infty \frac{W^2 + \left[\overline{w'^2} - \overline{v'^2}\right]}{\tilde{r}}d\tilde{r}\right\}.$$

$$(A.10)$$

Then, multiplying by r, assuming that the swirl velocity and turbulence second-order moments vanish at infinity, the integration in $[0, \infty)$ of equation (A.10) yields to:

$$\frac{d}{dx}\int_0^\infty \left[U^2 - \frac{W^2}{2} + \overline{u'^2} - \left(\frac{\overline{w'^2} + \overline{v'^2}}{2}\right)\right]r\,dr = 0. \quad (A.11)$$

Integrating equation (A.11) from the origin to any streamwise station yields the following streamwise momentum integral constraint on the motion:

$$M_0 = 2\pi \int_0^\infty \left[U^2 - \frac{W^2}{2} + \overline{u'^2} - \left(\frac{\overline{w'^2} + \overline{v'^2}}{2} \right) \right] r\, dr. \tag{A.12}$$

In order to obtain the angular momentum in axial direction the equation (A.3) has to be multiplied by W and then added to equation (A.6) which yields to:

$$\frac{\partial}{\partial x} UW + \frac{\partial}{\partial x} \overline{u'w'} = -\frac{1}{r^2} \frac{\partial}{\partial r} \left(r^2 VW \right) - \frac{1}{r^2} \frac{\partial}{\partial r} \left(r^2 \overline{v'w'} \right) \tag{A.13}$$

then multiplied to r^2

$$\frac{\partial}{\partial x} \left[r^2 \left(UW + \overline{u'w'} \right) \right] = -\frac{\partial}{\partial r} \left[r^2 \left(VW + \overline{v'w'} \right) \right]. \tag{A.14}$$

The final momentum integral is obtained by integrating equation (A.14) over r

$$\frac{d}{dx} \left[\int_0^\infty 2\pi \left(UW + \overline{u'w'} \right) r^2 dr \right] = - \left[2\pi r^2 \left(VW + \overline{v'w'} \right) \right]_0^\infty \tag{A.15}$$

with boundary conditions,

$$r = 0 \rightarrow V = W = \overline{u'v'} = 0 \tag{A.16}$$

$$r = \infty \rightarrow U = W = \overline{u'v'} = \overline{v'w'} = 0. \tag{A.17}$$

Hence, the right side of equation (A.15) vanishes:

$$\Rightarrow - \left[2\pi r^2 \left(VW + \overline{v'w'} \right) \right]_0^\infty = 0$$

and the left side of equation (A.15) yields:

$$\frac{d}{dx} \int_0^\infty 2\pi \left(UW + \overline{u'w'} \right) r^2 dr = 0. \tag{A.18}$$

Integrating, as before, from the origin to any streamwise location the equation (A.18) yields the second constraint to the flow motion which is the angular momentum integral:

$$G_\theta = 2\pi \int_0^\infty \left(UW + \overline{u'w'} \right) r^2 dr. \tag{A.19}$$

Hence, the two fundamental integrals reported in equations (A.12) and (A.19) represent the basic scaling parameters of a swirling motion. The first is the total rate of transfer of kinematic linear momentum across at generic location x. Since there are no net forces other than pressure which are taken into account in equation (A.12), M_x is constant and remain equal to its source value M_0 for all downstream positions $(M_x(x) = M_0)$.

The second parameter G_θ is the rate at which kinematic angular momentum is distributed across any downstream plane. Like the linear momentum, this parameter should remain constant at its source value G_0 since in an infinite environment there are no torques acting on any control volume containing the source plane. Therefore, a length scale can be defined

$$L^* = \frac{G_0}{M_0} \tag{A.20}$$

which can be referred to the nozzle radius $D/2$ yielding to

$$S = \frac{2G_0}{M_0 D} \tag{A.21}$$

which is the usual swirl number definition in case of top-hat velocity profiles.

B   Proper Orthogonal Decomposition

The availability of large datasets extracted from Direct Numerical Simulations (DNS) or from high spatial and temporal resolved snapshots acquired during experiments has the big potential to provide additional information in case of much complex flow phenomena. In this frame, the modal decomposition methods are efficient to reduce the dimensionality of a problem dividing the phenomena into basic components.

The Principal Component Analysis (PCA) was introduced to reduce dimensionality of large multi-dimensional data sets (Hotelling, 1933; Pearson, 1901). In the next years, different names were given depending on the field of application: Singular Values Decomposition (SVD), Karhunen–Loève (K-L) expansion and Proper Orthogonal Decomposition (POD). As a matter, they represent the solution to the same problem (Y. Liang et al., 2002) and the most recently accepted name is the last one mentioned.

One of the main advantages of POD is that it does not require to know the government equations of the phenomenon under investigation. This makes the modal decomposition suitable for a variety of fields, among them the study of turbulence. The first application of POD was the multivariate analysis of stochastic processes. Lumley (1967) demonstrated that the POD was objectively able to define coherent structures in turbulent flows. He showed that these can be detected as the most energetic spatial modes, thus ones which mostly contribute in variance to the acquired samples. For this reason, the POD spectrum is considered "optimum" because the variance in the POD-defined space decays faster than in any other vectorial space.

The practical implementation of a computationally efficient method for POD modes calculation was designed by Sirovich (1987). The method involved the use of a set of snapshots of the flow field. The modes are thus calculated as the projection of the flow field on the eigenfunctions of the temporal two-point correlation matrix. In this approach, the POD is used to identify a set of orthonormal functions which are the "optimum" to correlate with a given set of observations. As it is defined, the method, called *snapshot method*, is especially suited for field measurements provided by numerical simulations or Particle Image Velocimetry (PIV) snapshots. As a matter, both of them provide a large eulerian grid for a limited number of frames.

B.1 Mathematical framework

In the following mathematical treatment, adapted from Raiola (2017), the bi-orthogonal decomposition of Aubry et al. (1991) will be adopted. Hence, a large time sample of a dynamical system varying both in space and time will be considered. As stated before, the observations can be provided by experimental measurements or by a numerical solution of a generic vector field acquired at different time instants.

Let consider a vector field $\mathscr{F}(\underline{x}, t)$ approximated by the following discrete sum:

$$\mathscr{F}(\underline{x}, t) = \langle \mathscr{F}(\underline{x}, t) \rangle + \underline{f}(\underline{x}, t)$$
$$\approx \langle \mathscr{F}(\underline{x}, t) \rangle + \sum_{i=1}^{n_m} \psi^{(i)}(t) \lambda^{(i)} \underline{\phi}^{(i)}(\underline{x}) \quad \text{(B.22)}$$

where

- $\underline{x}$ and t are the space and time coordinates, respectively;

- the symbols $\langle \mathscr{F}(\underline{x}, t) \rangle$ and $\underline{f}(\underline{x}, t)$ indicate the ensemble average and the fluctuating part of the vector field $\mathscr{F}(\underline{x}, t)$ respectively;

- the functions $\underline{\phi}^{(i)}(\underline{x})$ represent the spatial decomposition basis of the fluctuating field;

- the functions $\psi^{(i)}(t)$ are the temporal basis of the fluctuating field;

- the quantities $\lambda^{(i)}$ are the norms associated at each spatio-temporal mode;

- n_m is the number of modes.

The equation (B.22) becomes an equality in case of $n_m \to \infty$. The solution of the decomposition reported in equation (B.22) depends on the chosen basis functions.

In the classical implementation by Lumley (1967) the POD decomposition technique aims to maximise the correlation between the vector field and the space domain functions $\underline{\phi}^{(i)}(\underline{x})$. In addition, the choice of these functions is limited to which with unitary norm

$$\left\langle \int_{\Omega_{\text{obs}}} \left(\underline{\phi}^{(i)}(\underline{x}), \underline{\phi}^{(j)}(\underline{x}) \right) d\underline{x} \right\rangle = \delta_{i,j} \tag{B.23}$$

where $(\cdot,\cdot)$ is the scalar product and $\delta_{i,j}$ is the Kronecker delta symbol. The solution to this problem is equivalent to solve the Fredholm equation:

$$\int_{\Omega_{\text{obs}}} C_s\left(\underline{x},\underline{x}'\right) \underline{\phi}^{(i)}\left(\underline{x}'\right) d\underline{x}' = \left|\lambda^{(i)}\right|^2 \underline{\phi}^{(i)}(\underline{x}) \tag{B.24}$$

where Ω_{obs} is the entire observation domain and $C_s\left(\underline{x},\underline{x}'\right)$ is the two-point spatial correlation matrix:

$$C_s\left(\underline{x},\underline{x}'\right) = \left\langle \left(\underline{f}\left(\underline{x},t\right), \underline{f}\left(\underline{x}',t\right) \right) \right\rangle. \tag{B.25}$$

An alternative implementation is the *snapshot method* provided by Sirovich (1987). In this case, the basis chosen maximises the projection of the vector field $\underline{f}(\underline{x},t)$ on the time domain functions $\psi^{(i)}(t)$. As in the previous case, the basis functions have to satisfy the orthogonality constraint:

$$\left\langle \psi^{(i)}(t), \psi^{(j)}(t) \right\rangle = \delta_{i,j}. \tag{B.26}$$

The resulting Fredholm equation for this problem is

$$\int_{T_{\text{obs}}} C_t\left(t,t'\right) \psi^{(i)}\left(t'\right) dt' = \left|\lambda^{(i)}\right|^2 \psi^{(i)}(t) \tag{B.27}$$

where T_{obs} is the observation time and $C_t\,(t, t')$ is the two-point temporal correlation matrix:

$$C_t(t, t') = \frac{\int_{\Omega_{\text{obs}}} \left(\underline{f}\,(\underline{x}, t)\,, \underline{f}\,(\underline{x}, t')\right) d\underline{x}}{\int_{\Omega_{\text{obs}}} d\underline{x}}. \tag{B.28}$$

The presented formulation does not adapt well to datasets produced by experiments and simulations because these kinds of data sets are in the discrete form of a spatial grid over a set of instantaneous realisations. The discrete form of POD formulation can be rewritten applying the matrix notation.

Suppose to have a dataset of n_t temporal realisations, each of them containing n_p measurement points. Each snapshot can be reshaped in a row vector $\underline{f}^{(j)} \in \mathbb{R}^{1 \times n_p}$. Hence, the dataset can be rearranged as a rectangular matrix $\underline{\underline{F}} \in \mathbb{R}^{n_t \times n_p}$ with rank $r \leq \min(n_t, n_p)$:

$$\underline{\underline{F}} = \begin{bmatrix} \underline{f}^{(1)} \\ \underline{f}^{(2)} \\ \vdots \\ \underline{f}^{(n_t)} \end{bmatrix} \in \mathbb{R}^{n_t \times n_p}. \tag{B.29}$$

The resulting Fredholm equation is equivalent to the eigenvalue problem of the two-point spatial correlation matrix of $\underline{\underline{F}}$ which is obtained by the product $\underline{\underline{F}}^T \underline{\underline{F}} \in \mathbb{R}^{n_p \times n_p}$:

$$\underline{\underline{F}}^T \underline{\underline{F}}\, \underline{\underline{\Phi}} = \underline{\underline{\Phi}}\, \underline{\underline{\Lambda}} \tag{B.30}$$

where $\underline{\underline{\Lambda}}$ is a diagonal matrix which contains the r eigenvalues in descending order, and $\underline{\underline{\Phi}} = \left[\underline{\phi}^{(1)}, \underline{\phi}^{(2)}, \ldots, \underline{\phi}^{(r)}\right]$ contains the eigenvectors of $\underline{\underline{F}}^T \underline{\underline{F}}$.

The *snapshot method* POD implementation results in the Fredholm equation which, differently from before, is equivalent to the eigenvalue problem of the two-point temporal correlation matrix of $\underline{\underline{F}}$ obtained by the product $\underline{\underline{F}}\,\underline{\underline{F}}^T \in \mathbb{R}^{n_t \times n_t}$:

$$\underline{\underline{F}}\,\underline{\underline{F}}^T\,\underline{\underline{\Psi}} = \underline{\underline{\Psi}}\,\underline{\underline{\Lambda}} \tag{B.31}$$

where $\underline{\underline{\Psi}} = \left[\underline{\psi}^{(1)}, \underline{\psi}^{(2)}, \ldots, \underline{\psi}^{(r)}\right]$ contains the eigenvectors of $\underline{\underline{F}}\,\underline{\underline{F}}^T$.

Since $\underline{\underline{F}}\,\underline{\underline{F}}^T$ and $\underline{\underline{F}}^T\underline{\underline{F}}$ are non-negative Hermitian matrices they have a set of r non-negative eigenvalues. Hence the solution of both equations (B.30) and (B.31) is:

$$\underline{\underline{F}} = \underline{\underline{\Psi}}\,\underline{\underline{\Lambda}}\,\underline{\underline{\Phi}}^T = \sum_{i=1}^{r} \underline{\psi}^{(i)} \lambda^{(i)} \underline{\phi}^{(i)T} \tag{B.32}$$

where $\underline{\underline{\Phi}}$ is the orthonormal basis for the rows of $\underline{\underline{F}}$, $\underline{\underline{\Psi}}$ is the orthonormal basis for the columns of $\underline{\underline{F}}$ and $\underline{\underline{\Lambda}}$ is a diagonal matrix containing the norm of each mode contribution.

C Extended Proper Orthogonal Decomposition

The wide application range of POD pushed the researches toward the investigation about new ways of applying such a modal decomposition technique. One of the most promising is the Extended POD (EPOD) firstly introduced by Maurel et al. (2001) and Borée (2003). The aim of this extension of POD method is to estimate the correlation between the spatial modes related to two different datasets which share a common temporal basis. As a matter, such a decomposition technique requires that the two datasets have to be simultaneously acquired, but the nature of spatial content can be different. In general it can happen that the two datasets have different sizes: $\underline{\underline{A}} \in \mathbb{R}^{n_t \times n_A}$ and $\underline{\underline{B}} \in \mathbb{R}^{n_t \times n_B}$.

The POD decomposition is still valid for both datasets, thus:

$$\underline{\underline{\Lambda}}_A \underline{\underline{\Phi}}_A^T = \underline{\underline{\Psi}}_A^T \underline{\underline{A}} \tag{C.33}$$

$$\underline{\underline{\Lambda}}_B \underline{\underline{\Phi}}_B^T = \underline{\underline{\Psi}}_B^T \underline{\underline{B}} \tag{C.34}$$

where

- $\underline{\underline{\Lambda}}_k$ is the diagonal matrix which contains the norms $\lambda_k^{(i)}$ of the projection of a dataset on the temporal modes of the same dataset;

- each column of $\underline{\underline{\Phi}}_k$ is the ith spatial mode of related snapshot matrix;

- each row of $\underline{\underline{\Psi}}_k$ is the ith temporal mode of related snapshot matrix.

In order to estimate the Extended POD modes the equation (C.33) can be rewritten as follows:

$$\underline{\underline{A}} = \underline{\underline{\Psi}}_A \underline{\underline{\Lambda}}_A \underline{\underline{\Phi}}_A^T. \tag{C.35}$$

Hence, from equation (C.34), assumed the synchronisation of the measurements, the $\underline{\underline{B}}$ dataset is projected on the temporal basis of the first dataset, i.e. $\underline{\underline{\Psi}}_A$, resulting in:

$$\underline{\underline{\Psi}}_A^T \, \underline{\underline{B}} = \underline{\underline{\Lambda}}_e \, \underline{\underline{\Phi}}_e^T \tag{C.36}$$

where $\underline{\underline{\Phi}}_e$ are the spatial modes which contain the correlated features of both datasets and $\underline{\underline{\Lambda}}_e$ is the related energetic content. It is worth noting that the modes contained in $\underline{\underline{\Phi}}_e$ are not in energy-ranked descending order as in the POD.

References

Abramovich, G. N., Girshovich, T., Krasheninnikov, S. I., Sekundov, A., & Smirnova, I. (1984). The theory of turbulent jets. *Moscow Izdatel Nauka*.

Abrantes, J. K., & Azevedo, L. (2006). Fluid flow characteristics of a swirl jet impinging on a flat plate. In *13th international symposium on applications of laser techniques to fluid mechanics, lisbon, portugal, june* (pp. 26–29). doi:10.1615/ihtc13.p16.230

Adrian, R. J., & Yao, C.-S. (1985). Pulsed laser technique application to liquid and gaseous flows and the scattering power of seed materials. *Applied optics*, *24*(1), 44–52. doi:10.1364/ao.24.000044

Afroz, F., & Sharif, M. A. (2018). Numerical study of turbulent annular impinging jet flow and heat transfer from a flat surface. *Applied Thermal Engineering*, *138*, 154–172. doi:10.1016/j.applthermaleng.2018.04.007

Ahmed, Z. U., Al-Abdeli, Y. M., & Guzzomi, F. G. (2015). Impingement pressure characteristics of swirling and non-swirling turbulent jets. *Experimental Thermal and Fluid Science*, *68*, 722–732. doi:10.1016/j.expthermflusci.2015.07.017

Alekseenko, S. V., Abdurakipov, S., Hrebtov, M. Y., Tokareva, M. P., Dulin, V. M., & Markovich, D. M. (2018). Coherent structures in the near-field of swirling turbulent jets: A tomographic piv study. *International Journal of Heat and Fluid Flow*, *70*, 363–379. doi:10.1016/j.ijheatfluidflow.2017.12.009

Alekseenko, S. V., Bilsky, A. V., Dulin, V. M., & Markovich, D. M. (2007). Experimental study of an impinging jet with different swirl rates. *International Journal of Heat and Fluid Flow*, *28*(6), 1340–1359. doi:10.1016/j.ijheatfluidflow.2007.05.011

Amini, Y., Mokhtari, M., Haghshenasfard, M., & Gerdroodbary, M. B. (2015). Heat transfer of swirling impinging jets ejected from nozzles with twisted tapes utilizing cfd technique. *Case Studies in Thermal Engineering*, *6*, 104–115. doi:10.1016/j.csite.2015.08.001

Antoranz, A., Ianiro, A., Flores, O., & García-Villalba, M. (2018). Extended proper orthogonal decomposition of non-homogeneous thermal fields in a turbulent pipe flow. *International Journal of Heat and Mass Transfer*, *118*, 1264–1275. doi:10.1016/j.ijheatmasstransfer.2017.11.076

Astarita, T., & Carlomagno, G. M. (2012). *Infrared thermography for thermo-fluid-dynamics*. doi:10.1007/978-3-642-29508-9

Atkinson, C., & Soria, J. (2009). An efficient simultaneous reconstruction technique for tomographic particle image velocimetry. *Experiments in Fluids*, *47*(4-5), 553. doi:10.1007/s00348-009-0728-0

Aubry, N., Guyonnet, R., & Lima, R. (1991). Spatiotemporal analysis of complex signals: Theory and applications. *Journal of Statistical Physics*, *64*(3-4), 683–739. doi:10.1007/bf01048312

Azevedo, L., Almeida, J., & Duarte, L. (1997). Mass transfer to swirling impinging jets. In *Proceedings of the 4th world conference on experimental heat transfer, fluid mech. and thermodynamics, brussels* (Vol. 3, pp. 1759–1766).

Bakirci, K., & Bilen, K. (2007). Visualization of heat transfer for impinging swirl flow. *Experimental thermal and fluid science*, *32*(1), 182–191. doi:10.1016/j.expthermflusci.2007.03.004

Ball, C., Fellouah, H., & Pollard, A. (2012). The flow field in turbulent round free jets. *Progress in Aerospace Sciences*, *50*, 1–26. doi:10.1016/j.paerosci.2011.10.002

Baughn, J. W., Hechanova, A. E., & Yan, X. (1991). An experimental study of entrainment effects on the heat transfer from a flat surface to a heated circular impinging jet. *Journal of Heat Transfer (Transactions of the ASME (American Society of Mechanical Engineers), Series C);(United States), 113*(4). doi:10.1115/1.2911197

Beér, J. M., & Chigier, N. A. (1972). Combustion aerodynamics. *New York*.

Benjamin, T. B. (1962). Theory of the vortex breakdown phenomenon. *Journal of Fluid Mechanics, 14*(4), 593–629. doi:10 . 1017 / s0022112062001482

Berkooz, G., Holmes, P., & Lumley, J. L. (1993). The proper orthogonal decomposition in the analysis of turbulent flows. *Annual review of fluid mechanics, 25*(1), 539–575. doi:10.1146/annurev.fluid.25.1.539

Bilen, K., Bakirci, K., Yapici, S., & Yavuz, T. (2002). Heat transfer from a plate impinging swirl jet. *International journal of energy research, 26*(4), 305–320. doi:10.1002/er.785

Billant, P., Chomaz, J.-M., & Huerre, P. (1998). Experimental study of vortex breakdown in swirling jets. *Journal of Fluid Mechanics, 376*, 183–219. doi:10.1017/s0022112098002870

Bolshukhin, M., Znamenskaya, I., & Fomichev, V. (2015). A method of quantitative analysis of rapid thermal processes through vessel walls under nonisothermal liquid flow. In *Doklady physics* (Vol. 60, *11*, pp. 524–527). Springer. doi:10.1134/s1028335815110014

Bolshukhin, M., Znamenskaya, I., Sveshnikov, D., & Fomichev, V. (2014). Thermographic study of turbulent water pulsations in nonisothermal mixing. *Optoelectronics, Instrumentation and Data Processing, 50*(5), 490–497. doi:10.3103/s8756699014050070

Borée, J. (2003). Extended proper orthogonal decomposition: A tool to analyse correlated events in turbulent flows. *Experiments in fluids, 35*(2), 188–192. doi:10.1007/s00348-003-0656-3

Bradshaw, P. (1966). The effect of initial conditions on the development of a free shear layer. *Journal of Fluid Mechanics, 26*(2), 225–236. doi:10.1017/s0022112066001204

Brown, K. J., Persoons, T., & Murray, D. B. (2010). Heat transfer characteristics of swirling impinging jets. In *2010 14th international heat transfer conference* (pp. 657 665). American Society of Mechanical Engineers. doi:10.1115/ihtc14-22860

Cala, C., Fernandes, E., Heitor, M., & Shtork, S. (2006). Coherent structures in unsteady swirling jet flow. *Experiments in Fluids, 40*(2), 267–276. doi:10.1007/s00348-005-0066-9

Candel, S., Durox, D., Schuller, T., Bourgouin, J.-F., & Moeck, J. P. (2014). Dynamics of swirling flames. *Annual review of fluid mechanics, 46*, 147–173. doi:10.1146/annurev-fluid-010313-141300

Carlomagno, G. M., & Ianiro, A. (2014). Thermo-fluid-dynamics of submerged jets impinging at short nozzle-to-plate distance: A review. *Experimental thermal and fluid science, 58*, 15–35. doi:10.1016/j.expthermflusci.2014.06.010

Chanaud, R. C. (1965). Observations of oscillatory motion in certain swirling flows. *Journal of Fluid Mechanics, 21*(1), 111–127. doi:10.1017/s0022112065000083

Chang, F., & Dhir, V. (1994). Turbulent flow field in tangentially injected swirl flows in tubes. *International Journal of Heat and Fluid Flow, 15*(5), 346–356. doi:10.1016/0142-727x(94)90048-5

Chigier, N. A., & Chervinsky, A. (1967). Experimental investigation of swirling vortex motion in jets. *Journal of Applied Mechanics, 34*(2), 443–451. doi:10.1115/1.3607703

Dahm, W., & Dimotakis, P. (1987). Measurements of entrainment and mixing in turbulent jets. *AIAA journal, 25*(9), 1216–1223. doi:10.2514/6.1985-56

Davidson, P. (2015). *Turbulence: An introduction for scientists and engineers.* doi:10.1063/1.2138427

Discetti, S. (2013). *Tomographic particle image velocimetry — developments and applications to turbulent flows* (Doctoral dissertation, Università degli Studi di Napoli "Federico II").

Discetti, S., & Astarita, T. (2010). Acceleration of tomo-piv by multigrid reconstruction schemes. In *15th international symposium on applications of laser techniques for fluid mechanics*.

Discetti, S., & Astarita, T. (2012). Fast 3d piv with direct sparse cross-correlations. *Experiments in fluids, 53*(5), 1437–1451. doi:10.1007/s00348-012-1370-9

Discetti, S., & Astarita, T. (2014). The detrimental effect of increasing the number of cameras on self-calibration for tomographic piv. *Measurement Science and Technology, 25*(8), 084001. doi:10.1088/0957-0233/25/8/084001

Dolling, D., & Gray, W. (1986). An experimental investigation of the mixing of coannular swirling flows. *AIAA journal, 24*(5), 785–792. doi:10.2514/3.9346

Eiamsa-ard, S., & Promvonge, P. (2005). Enhancement of heat transfer in a tube with regularly-spaced helical tape swirl generators. *Solar energy, 78*(4), 483–494. doi:10.1016/j.solener.2004.09.021

Elsinga, G. E. (2008). *Tomographic particle image velocimetry and its application to turbulent boundary layers* (Doctoral dissertation, TU Delft, Delft University of Technology).

Elsinga, G. E., Scarano, F., Wieneke, B., & van Oudheusden, B. W. (2006). Tomographic particle image velocimetry. *Experiments in fluids, 41*(6), 933–947. doi:10.1007/s00348-006-0212-z

Elsinga, G. E., Westerweel, J., Scarano, F., & Novara, M. (2011). On the velocity of ghost particles and the bias errors in tomographic-piv. *Experiments in fluids, 50*(4), 825–838. doi:10.1007/s00348-010-0930-0

Farokhi, S., Taghavi, R., & Rice, E. (1989). Effect of initial swirl distribution on the evolution of a turbulentjet. *AIAA journal, 27*(6), 700–706. doi:10.2514/3.10168

Fénot, M., Dorignac, E., & Lalizel, G. (2015). Heat transfer and flow structure of a multichannel impinging jet. *International Journal of Thermal Sciences*, *90*, 323–338. doi:10.1016/j.ijthermalsci.2014.12.006

Fiedler, H. (1988). Coherent structures in turbulent flows. *Progress in Aerospace Sciences*, *25*(3), 231–269. doi:10.1016/0376-0421(88)90001-2

Fudihara, T., Goldstein Jr, L., & Mori, M. (2003). The three-dimensional numerical aerodynamics of a movable block burner. *Brazilian Journal of Chemical Engineering*, *20*(4), 391–401. doi:10.1590/s0104-66322003000400006

Fudihara, T., Goldstein Jr, L., & Mori, M. (2007). A numerical investigation of the aerodynamics of a furnace with a movable block burner. *Brazilian Journal of Chemical Engineering*, *24*(2), 233–248. doi:10.1590/s0104-66322007000200008

Gauntner, J. W., Hrycak, P., & Livingood, J. (1970). Survey of literature on flow characteristics of a single turbulent jet impinging on a flat plate.

George, W. K. (1989). The self-preservation of turbulent flows and its relation to initial conditions and coherent structures. *Advances in turbulence*, *3973*.

Giaiotti, D., & Stel, F. (2006). The rankine vortex model. *October. University of Trieste*.

Gilchrist, R. T., & Naughton, J. W. (2005). Experimental study of incompressible jets with different initial swirl distributions: Mean results. *AIAA journal*, *43*(4), 741–751. doi:10.2514/1.3295

Goldstein, R., Sobolik, K., & Seol, W. (1990). Effect of entrainment on the heat transfer to a heated circular air jet impinging on a flat surface. *Journal of Heat Transfer*, *112*(3), 608–611. doi:10.1115/1.2910430

Gordon, R., Bender, R., & Herman, G. T. (1970). Algebraic reconstruction techniques (art) for three-dimensional electron microscopy and x-ray photography. *Journal of theoretical Biology, 29*(3), 471–481. doi:10.1016/0022-5193(70)90109-8

Görtler, H. (1954). Decay of swirl in an axially symmetrical jet, far from the orifice. *Revista matemática hispanoamericana, 14*(4), 143–178.

Graftieaux, L., Michard, M., & Grosjean, N. (2001). Combining piv, pod and vortex identification algorithms for the study of unsteady turbulent swirling flows. *Measurement Science and technology, 12*(9), 1422. doi:10.1088/0957-0233/12/9/307

Greco, C., Castrillo, G., Crispo, C., Astarita, T., & Cardone, G. (2016). Investigation of impinging single and twin circular synthetic jets flow field. *Experimental Thermal and Fluid Science, 74*, 354–367. doi:10.1016/j.expthermflusci.2015.12.019

Gupta, A. K., Lilley, D. G., & Syred, N. (1984). Swirl flows. *Tunbridge Wells, Kent, England, Abacus Press, 1984, 488 p.*

Haustein, H., Tebrügge, G., Rohlfs, W., & Kneer, R. (2012). Local heat transfer coefficient measurement through a visibly-transparent heater under jet-impingement cooling. *International Journal of Heat and Mass Transfer, 55*(23-24), 6410–6424. doi:10.1016/j.ijheatmasstransfer.2012.06.029

Hee, L. D., Youl, W. S., Taek, K. Y., & Suk, C. Y. (2002). Turbulent heat transfer from a flat surface to a swirling round impinging jet. *International Journal of Heat and Mass Transfer, 45*(1), 223–227. doi:10.1016/s0017-9310(01)00135-1

Herman, G. T., & Lent, A. (1976). Iterative reconstruction algorithms. *Computers in biology and medicine, 6*(4), 273–294. doi:10.1016/0010-4825(76)90066-4

Hohmann, S. (2011). Abstract: The confined tecflam swirling natural gas burner. Retrieved from http://www.kbwiki.ercoftac.org/w/index.php

Hotelling, H. (1933). Analysis of a complex of statistical variables into principal components. *Journal of educational psychology, 24*(6), 417. doi:10.1037/h0070888

Howell, J. R., Siegel, R., & Menguc, P. M. (2016). *Thermal radiation heat transfer*. CRC press.

Huang, L., & El-Genk, M. (1998). Heat transfer and flow visualization experiments of swirling, multi-channel, and conventional impinging jets. *International Journal of Heat and Mass Transfer, 41*(3), 583–600. doi:10.1016/s0017-9310(97)00123-3

Hunt, J. C., Wray, A. A., & Moin, P. (1988). Eddies, streams, and convergence zones in turbulent flows.

Hussain, F. (1986). Coherent structures and turbulence. *Journal of Fluid Mechanics, 173*, 303–356. doi:10.1017/S0022112086001192

Ianiro, A., & Cardone, G. (2012). Heat transfer rate and uniformity in multichannel swirling impinging jets. *Applied Thermal Engineering, 49*, 89–98. doi:10.1016/j.applthermaleng.2011.10.018

Ianiro, A., Lynch, K. P., Violato, D., Cardone, G., & Scarano, F. (2018). Three-dimensional organization and dynamics of vortices in multichannel swirling jets. *Journal of Fluid Mechanics, 843*, 180–210. doi:10.1017/jfm.2018.140

Jaafar, M. M., Jusoff, K., Osman, M. S., & Ishak, M. S. A. (2011). Combustor aerodynamic using radial swirler. *International Journal of Physical Sciences, 6*(13), 3091–3098. doi:10.5897/IJPS11.404

Jambunathan, K., Lai, E., Moss, M., & Button, B. (1992). A review of heat transfer data for single circular jet impingement. *International journal of heat and fluid flow, 13*(2), 106–115. doi:10.1016/0142-727X(92)90017-4

Kerrebrock, J. L. (1977). Small disturbances in turbomachine annuli with swirl. *AIAA Journal, 15*(6), 794–803. doi:10.2514/3.7370

Kinsella, C., Donnelly, B., O'Donovan, T., & Murray, D. (2008). Heat transfer enhancement from a horizontal surface by impinging swirl jets. In *5th european thermal-sciences conference* (p. 8).

Kurnia, J. C., Sasmito, A. P., Xu, P., & Mujumdar, A. S. (2017). Performance and potential energy saving of thermal dryer with intermittent impinging jet. *Applied Thermal Engineering, 113*, 246–258. doi:10.1016/j.applthermaleng.2016.11.036

Labus, T. L., & Symons, E. P. (1972). Experimental investigation of an axisymmetric free jet with an initially uniform velocity profile.

Landreth, C. C., & Adrian, R. J. (1990). Impingement of a low reynolds number turbulent circular jet onto a flat plate at normal incidence. *Experiments in Fluids, 9*(1-2), 74–84. doi:10.1007/bf00575338

Lee, S.-L. (1965). Axisymmetrical turbulent swirling jet. *Journal of Applied Mechanics, 32*(2), 258–262. doi:10.1115/1.3625793

Leibovich, S. (1978). The structure of vortex breakdown. *Annual review of fluid mechanics, 10*(1), 221–246. doi:10.1146/annurev.fl.10.010178.001253

Li, Y., Li, B., Qi, F., & Cheung, S. C. (2018). Flow and heat transfer of parallel multiple jets obliquely impinging on a flat surface. *Applied Thermal Engineering, 133*, 588–603. doi:10.1016/j.applthermaleng.2018.01.064

Liang, H., & Maxworthy, T. (2005). An experimental investigation of swirling jets. *Journal of Fluid Mechanics, 525*, 115–159. doi:10.1017/S0022112004002629

Liang, Y., Lee, H., Lim, S., Lin, W., Lee, K., & Wu, C. (2002). Proper orthogonal decomposition and its applications—part i: Theory. *Journal of Sound and vibration, 252*(3), 527–544. doi:10.1006/jsvi.2001.4041

Liepmann, D., & Gharib, M. (1992). The role of streamwise vorticity in the near-field entrainment of round jets. *Journal of Fluid Mechanics, 245*, 643–668. doi:10.1017/s0022112092000612

Lilley, D. G. (1974). Turbulent swirling flame prediction. *AIAA Journal, 12*(2), 219–223. doi:10.2514/3.49196

Lilley, D. G. (1977). Swirl flows in combustion: A review. *AIAA journal, 15*(8), 1063–1078. doi:10.2514/3.60756

List, E. (1982). Turbulent jets and plumes. *Annual review of fluid mechanics, 14*(1), 189–212. doi:10.1016/b978-0-08-026492-9.50005-0

Loitsyanskii, L. (1953). The propagation of a twisted jet in an unbounded space filled with the same fluid. *Prikladnaya Matematika i Mekhanika, 17*(1), 3–16.

Lumley, J. L. (1967). The structure of inhomogeneous turbulent flows. *Atmospheric turbulence and radio wave propagation.*

Lynch, K., & Scarano, F. (2015). An efficient and accurate approach to mte-mart for time-resolved tomographic piv. *Experiments in Fluids, 56*(3), 66. doi:10.1007/s00348-015-1934-6

Maas, H., Gruen, A., & Papantoniou, D. (1993). Particle tracking velocimetry in three-dimensional flows. *Experiments in Fluids, 15*(2), 133–146. doi:10.1007/bf00190953

Markovich, D., Abdurakipov, S., Chikishev, L., Dulin, V., & Hanjalić, K. (2014). Comparative analysis of low-and high-swirl confined flames and jets by proper orthogonal and dynamic mode decompositions. *Physics of Fluids, 26*(6), 065109. doi:10.1063/1.4884915

Martin, H. (1977). Heat and mass transfer between impinging gas jets and solid surfaces. In *Advances in heat transfer* (Vol. 13, pp. 1–60). doi:10.1016/s0065-2717(08)70221-1

Maurel, S., Borée, J., & Lumley, J. (2001). Extended proper orthogonal decomposition: Application to jet/vortex interaction. *Flow, Turbulence and Combustion, 67*(2), 125–136. doi:10.1023/A:1014050204350

McNaughton, K., & Sinclair, C. (1966). Submerged jets in short cylindrical flow vessels. *Journal of Fluid Mechanics, 25*(2), 367–375. doi:10.1017/s0022112066001708

Micklow, G. J., Roychoudhury, S., Nguyen, H. L., & Cline, M. C. (1993). Emissions reduction by varying the swirler airflow split in advanced gas turbine combustors. *Journal of engineering for gas turbines and power, 115*(3), 563–569. doi:10.1115/1.2906744

Moreira, R. G. (2001). Impingement drying of foods using hot air and superheated steam. *Journal of Food Engineering, 49*(4), 291–295. doi:10.1016/s0260-8774(00)00225-9

Mujumdar, A. S. (2014). *Impingement drying.* doi:10.1201/b17208-19

Novara, M., Batenburg, K. J., & Scarano, F. (2010). Motion tracking-enhanced mart for tomographic piv. *Measurement science and technology, 21*(3), 035401. doi:10.1088/0957-0233/21/3/035401

Nozaki, A., Igarashi, Y., & Hishida, K. (2003). Heat transfer mechanism of a swirling impinging jet in a stagnation region. *Heat Transfer—Asian Research: Co-sponsored by the Society of Chemical Engineers of Japan and the Heat Transfer Division of ASME, 32*(8), 663–673. doi:10.1002/htj.10120

Nuntadusit, C., Waehahyee, M., Bunyajitradulya, A., & Shakouchi, T. (2010). Heat transfer enhancement for a swirling jet impingement. In *14th international symposium on flow visualization (isfv14), exco daegu, korea.*

O'Donovan, T. S. (2005). Fluid flow and heat transfer of an impinging air jet. *University of Dublin.*

Pavlova, A., & Amitay, M. (2006). Electronic cooling using synthetic jet impingement. *Journal of heat transfer, 128*(9), 897–907. doi:10.1115/1.2241889

Pearson, K. (1901). Liii. on lines and planes of closest fit to systems of points in space. *The London, Edinburgh, and Dublin Philosophical Magazine and Journal of Science, 2*(11), 559–572. doi:10.1080/14786440109462720

Popiel, C. O., & Trass, O. (1991). Visualization of a free and impinging round jet. *Experimental Thermal and Fluid Science, 4*(3), 253–264. doi:10.1016/0894-1777(91)90043-q

Prahl, S. (2018). Optical absorption of water compendium. https://omlc.org/spectra/water/abs/index.html. [Online; accessed 25-June-2018].

Prandtl, L. (1904). Verhandlungen des dritten internationalen mathematiker-kongresses. *Heidelberg, Leipeizig,* 484–491.

Raffel, M., Willert, C. E., Scarano, F., Kähler, C. J., Wereley, S. T., & Kompenhans, J. (2018). *Particle image velocimetry: A practical guide*. Springer.

Raiola, M. (2017). *Empirical eigenfunctions: Application in unsteady aerodynamics* (Doctoral dissertation, Universidad Carlos III de Madrid).

Rajaratnam, N. (1976). *Turbulent jets*. doi:10.1017/S0022112077220866

Rohlfs, W., Haustein, H. D., Garbrecht, O., & Kneer, R. (2012). Insights into the local heat transfer of a submerged impinging jet: Influence of local flow acceleration and vortex-wall interaction. *International Journal of Heat and Mass Transfer*, *55*(25-26), 7728–7736. doi:10.1016/j.ijheatmasstransfer.2012.07.081

Rose, W. (1962). A swirling round turbulent jet: 1—mean-flow measurements. *Journal of Applied Mechanics*, *29*(4), 615–625. doi:10.1115/1.3640644

Sarpkaya, T. (1971). On stationary and travelling vortex breakdowns. *Journal of Fluid Mechanics*, *45*(3), 545–559. doi:10.1017/s0022112071000181

Scarano, F. (2001). Iterative image deformation methods in piv. *Measurement science and technology*, *13*(1), R1. doi:10.1088/0957-0233/13/1/201

Scarano, F. (2012). Tomographic piv: Principles and practice. *Measurement Science and Technology*, *24*(1), 012001. doi:10.1088/0957-0233/24/1/012001

Schanz, D., Gesemann, S., & Schröder, A. (2016). Shake-the-box: Lagrangian particle tracking at high particle image densities. *Experiments in fluids*, *57*(5), 70. doi:10.1007/s00348-016-2157-1

Schröder, A., Schanz, D., Michaelis, D., Cierpka, C., Scharnowski, S., & Kähler, C. (2015). Advances of piv and 4d-ptv "shake-the-box" for turbulent flow analysis–the flow over periodic hills. *Flow, Turbulence and Combustion*, *95*(2-3), 193–209. doi:10.1007/s10494-015-9616-2

Senda, M., Inaoka, K., Toyoda, D., & Sato, S. (2005). Heat transfer and fluid flow characteristics in a swirling impinging jet. *Heat Transfer—Asian Research: Co-sponsored by the Society of Chemical Engineers of Japan and the Heat Transfer Division of ASME, 34*(5), 324–335. doi:10.1002/htj.20068

Shiri, A. (2010). *Turbulence measurements in a natural convection boundary layer and a swirling jet* (Doctoral dissertation, Chalmers University of Technology).

Sirovich, L. (1987). Turbulence and the dynamics of coherent structures. i. coherent structures. *Quarterly of applied mathematics, 45*(3), 561–571. doi:10.1090/qam/910463

Smithberg, E., & Landis, F. (1964). Friction and forced convection heat-transfer characteristics in tubes with twisted tape swirl generators. *Journal of Heat Transfer, 86*(1), 39–48. doi:10.1115/1.3687062

Soloff, S. M., Adrian, R. J., & Liu, Z.-C. (1997). Distortion compensation for generalized stereoscopic particle image velocimetry. *Measurement science and technology, 8*(12), 1441. doi:10.1088/0957-0233/8/12/008

Steiger, M. H., & Bloom, M. H. (1962). Axially symmetric laminar free mixing with large swirl. *Journal of Heat Transfer, 84*(4), 370–374. doi:10.1115/1.3684400

Strang, E., & Fernando, H. (2001). Entrainment and mixing in stratified shear flows. *Journal of Fluid Mechanics, 428*, 349–386. doi:10.1017/s0022112000002706

Susan-Resiga, R., Muntean, S., & Bosioc, A. (2008). Blade design for swirling flow generator. In *Proceedings of the 4th german–romanian workshop on turbomachinery hydrodynamics (growth)*.

Syred, N., & Beer, J. (1974). Combustion in swirling flows: A review. *Combustion and flame, 23*(2), 143–201. doi:10.1016/0010-2180(74)90057-1

Syred, N., Chigier, N. A., & Beer, J. (1971). Flame stabilization in recirculation zones of jets with swirl. In *Symposium (international) on combustion* (Vol. 13, *1*, pp. 617–624). Elsevier. doi:10.1016/s0082-0784(71)80063-2

Syred, N. [Nicholas]. (2006). A review of oscillation mechanisms and the role of the precessing vortex core (pvc) in swirl combustion systems. *Progress in Energy and Combustion Science, 32*(2), 93–161. doi:10.1016/j.pecs.2005.10.002

Thianpong, C., Eiamsa-Ard, P., Wongcharee, K., & Eiamsa-Ard, S. (2009). Compound heat transfer enhancement of a dimpled tube with a twisted tape swirl generator. *International Communications in Heat and Mass Transfer, 36*(7), 698–704. doi:10.1016/j.icheatmasstransfer.2009.03.026

Thorlabs. (2018). Total transmission of 5mm thick, uncoated sapphire window. https : / / www . thorlabs . com / images / TabImages / WG31050_Raw_Data.xlsx. [Online; accessed 25-June-2018].

Toh, K., Honnery, D., & Soria, J. (2010). Axial plus tangential entry swirling jet. *Experiments in Fluids, 48*(2), 309–325. doi:10.1007/s00348-009-0734-2

Trüpel, A. W. T. (1914). *Über die einwirkung eines luftstrahles auf die umgebende luft.* Oldenbourg.

Tsai, R. (1987). A versatile camera calibration technique for high-accuracy 3d machine vision metrology using off-the-shelf tv cameras and lenses. *IEEE Journal on Robotics and Automation, 3*(4), 323–344. doi:10.1109/jra.1987.1087109

Tsukaguchi, Y., Nakamura, O., Jönsson, P., Yokoya, S., Tanaka, T., & Hara, S. (2007). Design of swirling flow submerged entry nozzles for optimal head consumption between tundish and mold. *ISIJ international, 47*(10), 1436–1443. doi:10.2355/isijinternational.47.1436

Violato, D., Ianiro, A., Cardone, G., & Scarano, F. (2012). Three-dimensional vortex dynamics and convective heat transfer in circular and chevron impinging jets. *International Journal of Heat and Fluid Flow, 37*, 22–36. doi:10.1016/j.ijheatfluidflow.2012.06.003

Violato, D., & Scarano, F. (2011). Three-dimensional evolution of flow structures in transitional circular and chevron jets. *Physics of Fluids*, *23*(12), 124104. doi:10.1063/1.3665141

Viskanta, R. (1993). Heat transfer to impinging isothermal gas and flame jets. *Experimental thermal and fluid science*, *6*(2), 111–134. doi:10.1016/0894-1777(93)90022-b

Ward, J., & Mahmood, M. (1982). Heat transfer from a turbulent, swirling, impinging jet. In *Heat transfer 1982, volume 3* (Vol. 3, pp. 401–407).

Warhaft, Z. (2000). Passive scalars in turbulent flows. *Annual Review of Fluid Mechanics*, *32*(1), 203–240. doi:10.1146/annurev.fluid.32.1.203

Wen, M.-Y., & Jang, K.-J. (2003). An impingement cooling on a flat surface by using circular jet with longitudinal swirling strips. *International Journal of Heat and Mass Transfer*, *46*(24), 4657–4667. doi:10.1016/s0017-9310(03)00302-8

Westerweel, J., & Scarano, F. (2005). Universal outlier detection for piv data. *Experiments in fluids*, *39*(6), 1096–1100. doi:10.1007/s00348-005-0016-6

Wieneke, B. (2008). Volume self-calibration for 3d particle image velocimetry. *Experiments in fluids*, *45*(4), 549–556. doi:10.1007/s00348-008-0521-5

Wieneke, B. (2012). Iterative reconstruction of volumetric particle distribution. *Measurement Science and Technology*, *24*(2), 024008. doi:10.1088/0957-0233/24/2/024008

Xu, G., & Antonia, R. (2002). Effect of different initial conditions on a turbulent round free jet. *Experiments in Fluids*, *33*(5), 677–683. doi:10.1007/s00348-002-0523-7

Yamada, S., & Nakamura, H. (2016). Construction of 2d-3c piv and high-speed infrared thermography combined system for simultaneous measurement of flow and thermal fluctuations over a backward facing step. *International Journal of Heat and Fluid Flow, 61*, 174–182. doi:10.1016/j.ijheatfluidflow.2016.04.010

Yule, A. (1978). Large-scale structure in the mixing layer of a round jet. *Journal of Fluid Mechanics, 89*(3), 413–432. doi:10.1017/s0022112078002670

Zeng, W., Sjöberg, M., & Reuss, D. L. (2015). Piv examination of spray-enhanced swirl flow for combustion stabilization in a spray-guided stratified-charge direct-injection spark-ignition engine. *International Journal of Engine Research, 16*(3), 306–322. doi:10.1177/1468087414564605

Znamenskaya, I., Koroteeva, E. Y., Novinskaya, A., & Sysoev, N. (2016a). High-speed ir thermography of submerged turbulent water jets. In *Proceedings of the 13th quantitative infrared thermography conference* (pp. 433–439). doi:10.21611/qirt.2016.063

Znamenskaya, I., Koroteeva, E. Y., Novinskaya, A., & Sysoev, N. (2016b). Spectral peculiarities of turbulent pulsations of submerged water jets. *Technical Physics Letters, 42*(7), 686–688. doi:10.1134/s1063785016070154

Zuckerman, N., & Lior, N. (2006). Jet impingement heat transfer: Physics, correlations, and numerical modeling. *Advances in heat transfer, 39*, 565–631. doi:10.1016/s0065-2717(06)39006-5

List of publications

Journal papers

- Raiola, M., Greco, C. S., Contino, M., Discetti, S., & Ianiro, A. (2017). Towards enabling time-resolved measurements of turbulent convective heat transfer maps with IR thermography and a heated thin foil, *International Journal of Heat and Mass Transfer*, 108, 199-209.
- Greco, C. S., Paolillo, G., Contino, M., Caramiello, C., Di Foggia, M., & Cardone, G. 3D temperature mapping of a ceramic shell mold in investment casting process via Infrared Thermography, *IEEE Transactions on Industrial Informatics*. Under review.
- Contino, M., Montanaro, A., Allocca, L., & Cardone, G. Experimental Infrared thermography investigation of heat transfer features of a multi-hole GDI spray. To be submitted.
- Contino, M., Greco, C. S., Ianiro, A., Astarita, T., & Cardone, G. Investigation of simultaneous thermal and velocity fields of a non-isothermal impinging swirling jet. To be submitted.

Conference contributions

- Contino, M., Greco, C. S., & Cardone, G. (2016). PIV measurements of sloshing flow in a rectangular tank induced by a sinusoidal inflow. 17th *International Symposium on flow Visualization.*

- Castrillo, G., Paolillo, G., Contino, M., Cafiero, G., & Astarita, T. (2016). Flow field features of fractal jets. In *Proceedings of the 18th International Symposium on the Application of Laser and Imaging Techniques to Fluid Mechanics.*

- Greco, C. S., Paolillo, G., Contino, M., Caramiello, C., Di Foggia, M., & Cardone, G. (2017). Investment casting process: 3D temperature map reconstruction of a ceramic shell mold. 9th *World Conference on Experimental Heat Transfer, Fluid Mechanics and Thermodynamics.*

- Contino, M., Montanaro, A., Allocca, L., & Cardone, G. (2017). IR thermography flow visualization of single and multi-hole spray impacting on a heated thin foil. 14th *International Conference on Fluid Control, Measurements and Visualization.*

- Contino, M., Montanaro, A., Allocca, L., & Cardone, G. (2018). Thermal investigation of a GDI multi-hole spray impact on a heated thin foil. In *Proceedings of the 3rd Thermal and Fluids Engineering Conference.*

- Contino, M., Greco, C. S., Astarita, T., & Cardone, G. (2018). Correlation between wall temperature and flow field of an impinging chevron jet. In *Proceedings of the 14th Quantitative InfraRed Thermography Conference.*

- Contino, M., Greco, C. S., Astarita, T., & Cardone, G. (2018). Simultaneous measurement of 3D velocity field and wall temperature distribution of an impinging chevron jet. In *Proceedings of the 5th International Conference on Experimental Fluid Mechanics.*

Acknowledgements

I think that this is the most difficult paragraph to be prepared, in particular because of the divergent emotions felt.

Typically, the feelings are different depending on who is writing, but the general sensation is that something is ending. Often, this is the kind of moment when you feel the need to sum up the path you took, the choices you made, the goals you reached. Each event you lived, each reproach or compliment you deserved, each person you met, the difficulties you encountered or the suggestions you received, everything takes part of you and makes you what you are at the end of this step.

It would be really difficult to list everything but I will give it a try to myself. At the end it would be like closing a book that entertained you for a long time, putting it on your chest, trying to feel everything once more.

I think that none of my goals would have been possible without my tutors: Prof. Gennaro Cardone and Prof. Carlo Salvatore Greco. I never saw Gennaro as my boss, rather like a friend. I always trusted him and I learned much. He often trusted me and I have always been honoured of that.

Carlo never refused my never ending requests for Skype calls, phone calls, meetings and WhatsApp chats to listen to worries, ideas and news that came up to my mind.

I shared my office with Carlo, but I never considered it a boring place thanks to Gerardo Paolillo, my PhD colleague. Thanks to his positive mind-oriented suggestions and to his presence, I never thought about the time I was spending at work. We shared much of the days of the last three years in the laboratory, endless source of contingencies (sometimes I thought that we were all in a dungeon's board game).

Many thanks also to Prof. Tommaso Astarita. He has been also my MSc tutor and I think that he lighted in me the passion for research. He has been always frank, he never sugarcoated the issues encountered, and I really appreciated it.

I think that Tommaso and me shared the luck of a few to be the front-line task force committed to fix the technology issues of Prof. Giovanni Maria Carlomagno. For the jobs done I earned an autographed copy of his book and I am really proud of that. He has been always a quiet presence, he always knew what to say in the right moment, often giving a simple solution to the problems.

I would thank all my PhD colleagues: Cuono Massimo Crispo taught me the importance of a "good morning" and of a "have a nice week-end" in a work place; Mario Ostieri shared with me many nerd hobbies and we always felt to be on the same wavelength; Giusy Castrillo has been always during all my BCs, MSc, and now, PhD studies; Simone Boccardi was always the happy part of the day.

I would also thank Giuseppe Sicardi, the laboratory technician who always tried, without succeeding, I admit for my fault, to teach me the importance of working without hurry during the manufacturing. He always also helped me to overcome and manage all annoying day-by-day bureaucracy.

The results and the goals reached in the present work would not have been possible without the help of Prof. Andrea Ianiro. He hosted me in Universidad Carlos III de Madrid for three months. I sincerely feel I learned as much as I was there for six months. He has the gift to see the good in all and he always provided me the right point of view to see the things. Andrea was not only my foreign tutor, but also the friend who taught me *un montón de cosas* about Madrid.

My house mates in Madrid were the key to my spare time. Hannah taught me some about Spanish language and Spanish cooking. Unfortunately, I never approved her way of seasoning a pizza. Nico taught me how important could be an *asado* for an Uruguayan guy. Christina taught me to take the job apart when the weekend is approaching. I really appreciated the long talks we had about everything that happened in the house.

My largest gratitude goes to my mother, tireless sink of my complaints, and to my father. I think he would be proud of me from up there. I would also thank my grandparents because of their unconditioned support.

My old and new friends, spread in Europe, always gave me the strength to go on, to fight, to rise again after the falls.

Well, I hope nobody has been forgotten. In case, I would apologise, but I am sure that even a word or a second shared with me has contributed to make me as now I am.

Mattia Contino